THE CRAFT
OF REVISION

THE CRAFT
OF REVISION

Second Edition

Donald M. Murray

Harcourt Brace College Publishers

Fort Worth   Philadelphia   San Diego
New York   Orlando   Austin   San Antonio
Toronto   Montreal   London   Sydney   Tokyo

Publisher	Ted Buchholz
Acquisitions Editor	Michael A. Rosenberg
Developmental Editor	Laurie Runion
Project Editor	Barbara Moreland
Production Manager	Serena B. Manning
Art Director	Sue Hart

ISBN: 0-15-501636-9

Library of Congress Catalog Card Number: 93-81254

Address for Editorial Correspondence: Harcourt Brace College Publishers, 301 Commerce Street, Suite 3700, Fort Worth, TX 76102.

Address for Orders: Harcourt Brace & Company, 6277 Sea Harbor Drive, Orlando, FL 32887. 1-800-782-4479, or 1-800-433-0001 (in Florida).

Printed in the United States of America

6 7 8 9 0 1 2 3   016   9 8 7 6 5 4

for Minnie Mae
Who made soup of old bones
and mailed out manuscripts
in which I had no faith

About the Author

 Donald M. Murray is Professor Emeritus of English at the University of New Hampshire. In addition to writing novels, poetry, and textbooks on writing and teaching writing, he also writes a weekly column "Over Sixty" for *The Boston Globe*. As a journalist, Murray has won a number of awards, including the Pulitzer Prize for editorial writing on the *Boston Herald* in 1954. In 1991, he was selected as the best columnist in Boston by *Boston* magazine. Murray was an editor of *Time* and a full-time writer in New York City before entering the academic world.

At the University of New Hampshire, Murray served as director of Freshman English and as English Department chairperson. He twice won awards for his teaching and was awarded honorary doctorate degrees from the University of New Hampshire and Fitchburg State College.

He has published more than a dozen books, including *Write to Learn*, Fourth Edition (Harcourt Brace College Publishers), *Read to Write*, Third Edition (Harcourt Brace College Publishers), *A Writer Teaches Writing*, Second Edition (Houghton Mifflin), *Learning by Teaching*, Second Edition (Heinemann, Boynton/Cook), *Writing for Your Readers*, Second Edition (Globe Pequot Press), *Expecting the Unexpected* (Heinemann, Boynton/Cook), and *Shoptalk* (Heinemann, Boynton/Cook).

PREFACE TO
THE SECOND EDITION

I practice what I preach. *The Craft of Revision, Second Edition* has been significantly revised.

In this book I argue that we build effective writing more from increasing the strengths of a text than from eliminating error. Frankly I was pleased with the first edition, more pleased than I have been with any other book I have written, but I was eager to rewrite it so I could increase its strengths.

Before I heard from my editor and my readers, I sat down and read the book, looking within my writing to see how it could be improved. In that reading, I heard the text and I heard the comments of many instructors and students, and I outlined the second edition.

Then I listened to my editor and my readers who had many valuable ideas and I incorporated them into my vision. Then I rewrote the entire book and, after it was read by my editor and more readers, I rewrote it again. Then I edited it.

All through this year-long process when I grumped about rewriting a book on revision, my wife, my daughters, and my friends laughed—and so did I. What did I expect?

A New Revision Process

My continuing research into the successful revision practice of publishing writers, advanced and beginning students, as well as close observation of my own evolving techniques, led me to reorganize the revision process in this edition. The new process is:

FOCUS The writer finds a limited territory to explore or a potential meaning to test by writing.

COLLECT The writer collects specific, accurate, revealing information.

SHAPE The writer chooses a genre and design that may carry the writer's meaning to the reader.

ORDER The writer lines the information up in a sequence that anticipates and answers the reader's questions.

DEVELOP The writer serves the reader a satisfying amount of information that makes the writer's meaning clear.

VOICE The writer tunes the language of the draft so that the music of the words support and reinforce the meaning.

EDIT The writer becomes the reader's advocate and makes sure that any exceptions to tradition clarify the meaning.

Of course the writing process of the individual student will change depending on the specific writing task, the experience of the writer with that task, the cognitive style of the writer, and the considerations imposed by teacher or editor. This new emphasis, however, will help students find a focus more quickly and develop it more efficiently.

A New Chapter on Finding a Subject

We have added a new first chapter to help students find a subject. To revise, students first have to find a subject and explore its possibilities. We have given students more strategies for exploring their personal and academic worlds and for discovering the writing possibilities in them.

A New Chapter on Reading

Reading what is—and what is not yet—written is such an essential writing and rewriting skill that we have devoted the second chapter to this subject. The student is presented with many reading techniques and is shown how one professional writer reads his own work.

New Case Histories

The second edition of *The Craft of Revision* now has five case histories that show students revising with earlier and later drafts that can be used for student examination and class discussion. This edition also has three professional case histories and far more examples of writing in process woven through the text than in the first edition.

New Focus on Diagnosis

Most beginning students learn to write and revise most effectively through working on papers based on personal experience, but they are students, and the purpose of the majority of writing courses is to prepare students to write within the academy. One of the strands woven through this edition is material that relates each stage of the revision process to academic writing.

Although *The Craft of Revision* is constructed on a firm theoretical base, my attempt has been to be practical at each stage of the rewriting process. During my years in the classroom I wanted materials that answered the pragmatic needs of my students and, as a writer who goes to his writing desk 365 days a year, I wanted writing counsel that can be applied directly to the pace under my hand. I hope I supply that to my readers.

Acknowledgments

Each book is a collaboration and many names should be listed on the title page. Laurie Runion, one of the very best editors with whom I have ever worked, has made each page of this edition better because of her perceptive questions and suggestions. She has the unusual ability to see the specific in the context of the whole. Michael Rosenberg saved the first edition from total failure. Without him, there would have been no first edition to revise.

My wife, Minnie Mae, is my first reader and most demanding critic. Chip Scanlan, my closest writing friend, is always as close as the telephone.

Professor Lisa Miller, a close friend and expert reader of many of my writings, was especially helpful in supplying all but one of the student case histories in this edition. Roger LePage made this a far better book with the case history of his essay and especially with his careful response to my editing of his writing.

Dr. Brock Dethier, Dr. Thomas Newkirk, Dr. Donald Graves and many other colleagues in the writing community at the University of New Hampshire have contributed to my continuing education. Others who have stimulated my thinking and influenced this book include Dr. Bonnie Sunstein of the University of Iowa, Dr. Thomas Romano and Dr. William Strong of Utah State University, Dr. Driek Zirinsky of Boise State University, Dr. Lad Tobin of Boston College, the writer, Ralph Fletcher, and my *Boston Globe* editor, Evelynne Kramer, who teaches me revision 52 weeks a year.

Dr. Mary Clark of the English Department at the University of New Hampshire is an expert linguist and a fine teacher who made important contributions to the first edition and her mark remains on this edition.

I am indebted to the professionalism of Barbara Moreland, project editor; Katherine Lincoln, copyeditor; and Sue Hart, designer.

The reviewers who helped me with their candid and insightful comments on this edition are: Kay Baker, Ricks College; Mary Comstock, University of Puget Sound; Francine DeFrance, Cerritos College; Connie Hale, University of Puget Sound; Dick Harrington, Piedmont Virginia Community College; Pat Huyett, University of Missouri at Kansas City; Leslie Prast, Delta College; David Roberts, Samford University; Susan Roberts, Boston College; Bernard Selzler, University of Minnesota at Crookston; and Nancy Walker, Southwest Missouri State University.

I was also instructed by the honest, detailed responses of those who reviewed the first edition: Kathleen Bell, Old Dominion University; Marie Czarnecki, Mohawk Valley Community College; Ernest Lee, Carson-Newman College; Joan Tyler Mead, Marshall University; and Driek Zirinsky, Boise State University.

PREFACE TO
THE FIRST EDITION

"Revise," we command, and our students change some of the punctuation, often trading new grammatical errors for old; choose a couple of long words they don't really know from Roget to "profound it up" as one of my students said; misspell a number of words in a more innovative way; catch a few typos, and pass back essentially the same paper.

It is all they know.

And it was all I knew until, after I had worked for newspapers for six years, I learned the craft of magazine and book writing. The first work I did in studying the writing process focused on revision; now, years later, I have returned to the craft of revision.

▪ WRITING IS REWRITING ▪

At our desks, writers make writing work. We take the rare gifts of inspiration, usually as wonderful and ugly as a new-born baby, and nurture them so they grow into essay, poem, novel, textbook, review, report, whatever is their destiny. And when we don't have inspiration, which is most of the time, we know how to produce a first draft we can work—by revision—into a final, publishable draft.

We know that revision is not punishment; that writing evolves from a sequence of drafts, each one teaching the writer how to write the next one. And we know that revision is copy editing and much, much more.

Scientists know that failure is essential, central, and necessary to their trade. They experiment and from the experiments that don't work—most of them—they discover the questions they need to ask, the method of asking them, and eventually, some answers. They revise.

Actors and musicians rehearse. Retailers test markets, politicians take polls, manufacturers try pilot runs. They all revise, and so do writers. *Writing is rewriting.*

■ THE CRAFT OF REVISION ■

Writing is a decision-making process. As we revise, considering each word, each piece of punctuation, each phrase, sentence, paragraph, page, we make decisions that lead to other decisions. We don't work by intuition but by craft.

This text takes the student into the workshop—into the head—of one publishing writer while writing is being made through revision. There are differences among the way writers work and this is taken into account, based on my study of how writers write that I began in junior high school. The text reveals the attitudes and the skills of the revising writer.

It is interesting that both good students and poor ones are equally unable to revise. The best students fall in love with their first drafts, and when they make them superficially correct they think they are finished; the poorest students also look to the mysteries of presentation—spelling, handwriting, mechanics, usage—and see nothing beyond that. But revision is based on re-seeing the entire piece of writing.

This book shows the student how to create a discovery draft and then how to read the draft to see what has been discovered. Then the text takes the student through the decisions of revision: focus, audience, form, information, structure, and language. They are dealt with in an orderly, sequential manner that the student can adapt to a variety of writing tasks. The student has a writer companion who helps the student read and revise the student's own draft until it communicates the draft's own meaning with clarity and grace.

▪ WHAT WILL MAKE STUDENTS WANT TO REVISE? ▪

The writer's motivation to revise comes from a positive and a negative force—and so will the student's. The positive force is the surprise of discovery. Writers are born at the moment they write what they do not expect and find a potential significance in what is on the page. They add a word or two, cut a few, move the others around, and watch that potential meaning come clear. They are hooked because the act of writing that, in the past, had revealed their ignorance, now reveals that they know more than they thought they knew, had more to say than they had realized.

Now they can understand what Hemingway meant when he said, "Prose is architecture not interior decoration." The students had thought that revising was only a superficial—to them—matter of etiquette and neatness, a dressing up of a message to which they had no commitment. Now they have a message they want heard and understood by readers. They are ready to be motivated by the negative force of reader understanding.

Students recognize that they need the craft of clarity and the craft of grace. To be read and understood they must know the traditions of language and break them only when it results in increased clarity. They have to meet the reader's expectations and find a common ground on which they can communicate.

Rewriting is, above all, a matter of attitude. And the teacher must model an attitude that emphasizes discovery and then communication. The teacher who says, "We all hate to rewrite but . . ." will breed a class that hates rewriting. The teacher who knows firsthand the excitement of revision, and may even share examples of personal revision, may make an interest in revision contagious.

■ HOW CAN STUDENTS LEARN THE CRAFT OF REVISION? ■

Students, first of all, must learn a positive attitude toward revision, which has usually been taught as punishment. The process of revision, for most students, has not been concerned with finding meaning, but has focused on editing superficial mechanical and grammatical errors to a preconceived and often not clearly understood standard. It is important for students to see revision in a larger context that *includes* editing, but is *not only* a matter of editing.

Some ways that the students' context can be expanded in the composition class include:

- First readings by the student writer, the instructor, and classmates should focus on potential, not error.
- There should be time for revision. Usually this means fewer papers, revised more extensively.
- The best as well as the worst papers should have the benefits of revision.
- Students should observe the discoveries of meaning made clear by revision on classmates' papers as well as their own.
- The process of revision should be sequential, moving from a concern with meaning through audience, form, information, structure, to language.

Above all, attitude motivates the learning of skills. The instructor should reinforce a constructive attitude toward revision. Some of the ways this can be done are:

- Share evolving drafts that document the positive results of revision. Reveal the instructor's own revision case histories, have students who have revised effectively share their own case histories, share the case histories of successful revision from interviews and biographies of writers.
- Have class members and possibly faculty members from other disciplines report on activities in their disciplines that are

similar to revision: dramatic and musical rehearsal, practice in a sport, the process of painting, experiments in science.

- Have the class perform quick revisions, for example, writing a five-line description of a familiar place or person in five minutes, then do five-minute rewrites from a different point of view, for a different purpose, for a different audience, in a different form, in a different voice, and share the results so the class appreciates the diverse products of revision.

At each stage of the revision process, students should play with the skills of shifting focus, appealing to different audiences, experimenting with form, manipulating information, restructuring, turning the draft's voice.

■ SIX WAYS TO USE THE CRAFT OF REVISION IN YOUR CLASSROOM ■

This text has been designed to support the student and instructor in many different courses. It can be used alone or as a supplement to other rhetorics and readers. Teachers will and should find their own ways to adapt this text to their particular teaching style, the needs of their own students, and the curriculum in which they function. These suggestions are designed to spark the diversity that should be central to teaching writing.

1. *The Craft of Revision* can be used as the principal writing text since it helps the student create a first draft. The students can use it to help them write weekly papers, or it can be used to support a sequence of three-week units in which the student writes a draft and performs two major revisions supported by conferences, peer workshops, and class instruction sessions. I have had good results with students working on one paper all semester, taking several weeks to find the subject, then moving week-by-week through the revision process. They didn't get bored because they found subjects they wanted to explore. In fact, class-alumni response has been unusually strong, saying they really learned to write when they had time to learn the craft of revision.

2. This text can supplement a rhetoric that has limited material on revision, or it can supplement a reader, allowing the students to understand the craft that created the models and to practice the same craft on their own drafts.

3. *The Craft of Revision* can be introduced to the class in the first week or two and then used as a reference book by the students as they revise their papers. The best way to do this is to have the students write a paragraph in class, then revise it a number of times, sharing each revision with a small group who will select the most interesting leap towards meaning to share with the whole class.

The students may need to be told in the beginning what to do, since they have only been told to edit in the past. I usually say, "Develop the potential in your draft, exploring the subject in writing any way you want, but if you're stuck you may want to change the point of view from which the subject is seen."

Some other suggestions I may make are: try to make the information more specific, revise it for a different publication or audience, write with more emotion or less, try to write in a different voice.

The writing periods should be short, no more than five minutes, and the peer sharing session no longer than fifteen minutes. It always helps if the instructor performs the same exercise and shares those drafts. Once the students experience the discovery that is possible with revision, they can be introduced to the text.

4. The text may be used in the latter half of a composition semester or term when the students have drafts worthy of careful revision and when they see the need for revision.

5. The text may be used in a content course in English or any other discipline to help the students improve their writing assignments. The text may be assigned to all students in the course or suggested as an aid to those students who are having difficulty revising effectively.

6. *The Craft of Revision* can be used as a self-teaching text with the student creating a draft and moving through the sequential steps to practice the skills of revision.

In every case the students should use the text in connection with their own writing. The craft of revision cannot be learned in the abstract; theory must be illuminated by practice that will, in turn, illuminate theory.

And however the text is used it will help the individual student, the instructor, and the whole class if those students who do an effective job of revision testify to the class on what they did and how. Sharing will reinforce the student and instruct everyone else. I have done this with oral reports but more recently with quickly written but complete commentaries that the students write, reporting on their writing and revision process, their writing problems and their proposed solutions. These commentaries encourage students to examine their craft and teach the student how they can identify and solve their own writing problems. And these solutions instruct us all.

My students have always instructed me, and I have told worried, inexperienced writing teachers to get their students writing. In every class, some students write better than others. Get them to tell you and the class how they wrote and the curriculum will evolve.

CONTENTS

■════════■

CHAPTER 1

Write to Re-Write   1

CHAPTER 2

Read to Re-Write   21

CHAPTER 7

Re-Write to Order   125

CHAPTER 8

Re-Write to Develop   142

CHAPTER 9

Re-Write with Voice   168

CHAPTER 1
WRITE TO RE-WRITE

You hear the command: "Write!"

You panic. You have nothing to say. Your head feels empty, the inside of a silent bass drum. Black. Nothing. No thoughts. No opinions. No memories. No ideas. No feelings. Nothing.

Good.

That's the terrifying place where the best writing begins.

I know that as a publishing writer, but I have to keep reminding myself it is true because the terror is real.

This is what I do when "emptyitis" strikes.

■ HOW DO I FIND SOMETHING TO WRITE ABOUT? ■

I relax and remind myself that this terrifying emptiness is where good writing comes from. I interview myself:

- What am I thinking about when I'm waiting for somebody?
- What irritated me today?
- What made me laugh?
- What made me angry?
- What did I learn today?
- What contradicted what I know—or thought I knew?

1

- What made me feel good?
- What made me feel bad?
- What confuses me?
- What does somebody else need to know that I know?
- What questions do I need answered?
- What surprised me today?

Try it. Write down your fragmentary answers. See if they connect. Follow one in your mind or on paper to see where your thinking may go.

Find the Instigating Line or Image

Inexperienced writers believe that writing begins with an inflated idea of a vague, general topic such as "truth," "beauty," or "patriotism" because they have been given such assignments in school. What they don't know is that published writers would do as badly as they do with such assignments unless they can come up with a line or an image that is specific and, above all, interesting to them.

A line is a fragment of language, a sentence or less, that I hear in my mind or find myself scribbling in my notebook. The line that contains a tension, contradiction, question, feeling, or thought that surprises me and would be productive to think more about in writing.

Let's take those vague topics I mentioned above and see what would happen if we had an instigating line to start us writing:

"truth"

"My first football coach was a priest—taught us how to lie—to fake injuries."

The conflict between the religious practice and football practice interests me. In church we were instructed to tell the truth and on the playing field we practiced fake injuries that would give us a time-out that we didn't deserve—and an unfair advantage over the other team.

I gave a lecture at a college whose football team was called "The Fighting Quakers," a good example of an instigating line as we imagine a group of hard-hitting pacifists. There are many essays in exploring the messages given by sports that contradict other social messages such as "don't fight," "winning isn't everything," "violence is bad."

"beauty"

"told mother I was fat, guys made jokes about me. She had same thing happen at same school—she was too thin, no 'sweater' girl. She weighed same as I weigh."

This observation presents a conflict that needs exploration. Women are shaped by nature and so are men. Fat and thin are relative terms. So is "beauty." Look at the pictures of film stars years ago. How dangerous it is to allow society to define you. How many are hurt who are not considered beautiful? How many are hurt by being considered a beauty (or a man who looks like a "hunk") and not taken seriously as a student? Those are a few of the topics that might grow from one line.

"patriotism"

"spies are traitorous patriots"

A fascinating and thoughtful essay for a course in history, ethics or political science could grow out of this fragmentary idea. Graham Greene once asked, "Isn't disloyalty as much the writer's virtue as loyalty is the soldier's?" Good question. Is the role of artists to stand apart from society and take stock? What about priests who criticize the Vatican, senators who vote against their party, soldiers who oppose a particular war? And what is a patriot anyway? Irish men and women are patriots in Ireland and traitors in Northern Ireland—and the other way around. All these are good pieces that could be researched and written.

Writers also find writing beginning for them with an image, a mind picture that itches, that makes you look at it again and again to see what it means.

All of us have images that haunt us. I remember seeing my face when I was a baby reflected in the glass of a china cabinet and that became a poem. I remember the long, empty corridors of high school between classes, so different from the way I usually saw them crowded with students. When I found why I remembered them— why I was so often alone in the corridor—I found an essay.

When I am given an assignment to write an academic paper, I do the same thing I do when searching for a poem. I sit with pen in hand and notebook open or without anything but my own thoughts. Sometimes I think about what I think about when I am not thinking: when I'm waiting for class to begin, sitting in a car waiting for someone, when my mind drifts away from the people to whom I should be listening, during the commercials on television, when I'm walking alone, just before I drop off to sleep. At these times, an image, a word, a fragment of language will pass through the black emptiness like a shooting star. I capture it in a scribbled note.

Other times I listen to what I'm saying when I talk to myself, or I remember what has surprised me recently. What did I see, think, feel, hear, watch, read that was not expected—in fact that ran against expectations, shocking or confusing me—contradicting what I thought I had known or believed?

Many people believe writing comes to the writer like a computer printout, flowing along, finished, complete. Writing usually comes in fragments—details, hints, clues, collisions of information, half ideas and quarter ideas, bits of pieces of information, scraps that have fallen out of books, from TV or radio, from conversations at the next table or in another room. The writer plays with these scraps to see what they may mean.

I start listing in a notebook or on a computer what passes through my unthinking mind. This technique is called **brainstorming**. It is a method of extracting from memory what you don't know you remember, what you have forgotten, or what you were never aware you observed, and allowing these pieces of information to rub against one another. Brainstorming works with ideas as well as memory, or with information collected for an academic paper.

To brainstorm, put aside all your notes, take a piece of paper and put down whatever occurs to you in a fragmentary list. Surprise yourself. Be silly, dumb—an enemy to your own preconceived ideas. You can work by yourself or with a team. Brainstorming is ideal when a committee is planning a party, a marketing campaign, a new research project. Work fast, with as little conscious thought as possible. When you are done, circle the items that surprised you the most and draw arrows between items that have some connection. This should produce an instigating line.

I begin by just thinking about a time in my life—high school— and start brainstorming:

The Principal, *Mister* Collins

North Quincy High

Long prison corridors, lockers, students locked in cells—classrooms

Round glasses on round face

Can't see his eyes

Cold oak bench outside principal's office

Detention

Boring

Not in school when in school

Working in meat market

Sitting in class, eyes open, asleep

Circle the Line or Image That Surprises

I suspend my critical judgment, not looking for good writing, but simply a word, a phrase, an image, a hint of an idea, two items or three that connect or are in conflict—whatever surprises or interests me.

Surprise is the most productive reaction for me; I like to discover that I have written something that itches, something I would not have predicted I would have put on my list.

The surprise doesn't have to be a history-shattering idea, just something that I hadn't thought about—not in that way anyway. I look at one item on my list—*"not in school when in school"*—and ask myself what that means. I slept and ate—mostly—at home, and I attended school at least four days a week. But I lived on the street. The gang on the street corner was the family that counted most. My real and imagined lives were at work and on the street, not in my flat or in my classrooms. No surprise there. I cut school a lot—every Thursday my senior year—and dropped out twice. No surprise to me in that. I have written that story before. But I had never faced up to the fact that I was *"not in school when in school."*

Looking back, that interests me. I don't know if I have anything to say, but I may. Writing may tell me. That phrase—*"not in school when in school"*—is a small surprise, but it sparks questions: Where was I? What did I find there that I didn't find in school? How did I manage to be in two places at once? And *"sitting in class, eyes open, asleep."* Well, that's one form of escape. How did I do that?

Explore the Surprising Line or Image

But I don't start writing yet. I make another list focussing on the item or items I've chosen to explore—*"sitting, eyes open, asleep"; "not in school when in school"*—and again list anything that comes into my mind that is connected to what I am exploring. I scribble specific details, statistics, facts, memories, observations, feelings, fragmentary thoughts—not in sentences but in words and small chunks of language.

> marching asleep in the Army
> looking awake but sleeping in faculty meetings
> during commercials
> at family dinners
> escaping into fantasy
> > memory
> > stories I tell myself
> > other places
> > other people—"empathy"

protection against bores
> against boring sermons
> classes
> jobs

how I look as if I'm listening when I'm not

how can I tell someone else is not listening to me when they look as if they are

———————————— ▪ *Writing Exercise* ▪ ————————————

Now it is time for you to check out what I have said to see if it works for you.

Take a piece of paper—or an empty computer screen—and brainstorm, writing down whatever passes through your mind. Don't worry about sentences, but be as specific as you can in a few words: not *"high prison fence"* but *"prison fence with coils of razor-edged steel at the top";* not *"could see road"* but *"on the road I saw people going and coming—nothing special to me until I could not go or come, just serve my sentence."*

Write fast to catch the thoughts, memories, ideals, specific details, you didn't know were in your mind. Five minutes is fine, ten good, fifteen probably too long. Stop and read the list, circling what surprised you and connecting with line and arrow those items that are related. Then list the topics you have discovered that you need to explore in writing.

▪ HOW DO I CREATE A DISCOVERY DRAFT? ▪

Now it is time to explore the subject further in a first or rough draft that I call a **discovery draft**. Once more, I do not write what I already know or what I expect to say but I write what I do *not* know. I am thinking in writing, the most disciplined form of thought. And I find it fun because I keep finding I know more than I expected, feel more than I expected, remember more and have a stronger opinion than I expected.

Write with Velocity

I write as fast as I can because velocity is as important in writing a discovery draft as it is in riding a bicycle. You have to get up to speed to get anywhere. My handwriting and my typing are appallingly bad. It is hard to read what I've written at top speed but I must write with velocity to:

Outrace the Censor

All of us have parents, teachers, editors, the critical self that makes us distrust ourselves—that censor what we say. It is important to write as fast as possible, not worrying about making sense or following the rules of language so that we can discover what has been hidden in our minds.

Fail

You mean you want failure? Yes. On the discovery draft, failure is essential. My best writing comes when I cannot yet understand what I am saying but find in it something to think about in further writing. The best touches of style occur when I say badly what I cannot yet say clearly and gracefully. When I understand the thought, the language will come that will reveal it because I have had to think in language.

No failures, no **instructive failures**—the accidents of insight and language—are in vain, as these tell us what direction we need to explore and develop.

Speed makes you say what you did not expect to say and reveals what you didn't know you knew.

Speed causes accidents. That's not good on a bike but it's good in writing a discovery draft. In fact, it's essential. Writing fast causes the instructive accidents of thought and feeling that produce a draft worthy of revision.

Write Out Loud

We learned to speak before we wrote and, even if we are writers, we speak thousands upon thousands of words more than we write in a day. When we write, we speak in written words.

And the magic of writing is that readers who never meet us, hear what we have written. Music rises from the page when we read.

We call the heard quality of writing **voice** and it may be the most important element in writing. Voice, like background music in a movie, is tuned to the text, supports and extends what the text says. Listen to master writers speak of voice:

Imagine yourself at your kitchen table, in your pajamas. Imagine one person you'd allow to see you that way, and write in the voice you'd use to that friend. SANDRA CISNEROS

I read everything that I write aloud. First, the paragraph. Then, the page. Then, the chapter. And finally, I read the whole book aloud. Because I want to hear my voice reading it, and I need it to sound natural. ISABEL ALLENDE

I hear a thousand "voices" in my head. They are the voices of my characters. They are male, female, white, brown and yellow. BHARATI MUKHERJEE

The longer you stay a writer, the more voices you find in your own voice and the more voices you find in the world. ALLAN GURGANUS

But voice serves the writer before the reader. It is the voice of the text that tells the writer the potential meaning of what is written in a discovery draft, reveals its emotional intensity, its importance to the writer and to readers. A college freshman might write:

My first week-end I expected everything to be the same as I left. The house had shrunk, the driveway was shorter than I remembered. The yard was smaller and my room had become a sewing room. My girl friend had become a friend, a distant friend who asked me for advice about some Elmer. I gave it. As if I was some university stud, not a guy who hadn't even walked to class with a girl yet. My old friends were gone and my folks treated me like company. I was a stranger to my life and I liked it. I wasn't that kid that lived in my room. Mother came home early from the real estate office and we just talked as if I were her son's friend home from college. When dad came home we had a beer and we went out to dinner with Uncle Val and my latest Aunt. They treated me as if I were grown up, dirty jokes and everything, and perhaps I was, perhaps my folks and I were friends.

I hear an individual voice, the voice of a person I'd like to know, someone who experiences the unexpected and makes something of it. I like the music of *"The house had shrunk, the driveway was shorter than I remembered," "some Elmer,"* and *"I was a stranger to my life and I liked it."* I suspect that voice has not yet been heard by the writer but it should be.

That draft needs to be developed, but it is a good first draft—a voice can be heard in a tailgating car wreck of language.

In the next draft, a student assigned to write a paper on George Orwell's essay "The Hanging" begins with a rough discovery draft:

I didn't like being assigned the Orwell essay in freshman English. I felt the instructor was messing with my head. My father is a sheriff and we believe in capital punishment and this is an English class and not political science or Liberal I. But I had to write a critical paper and read the essay one more time. It seemed to be about standing by and not doing anything like when Joe was being made fun of. I didn't join in but I didn't make friends either. After his suicide we tried to understand, I tried to, if there was anything I—we—could have done. I remembered seeing what was happening, not knowing what would happen of course, at a distance. Orwell knew what was happening but he was a colonial officer, part of the system, and so was I, I suppose, so was I. Standing, watching, doing nothing. That coldness was scary and I guess, it is an English paper, see how Orwell did that.

She has surprised herself by her second reading, by the connection with an incident in high school, and by her voice that recreates her guilt at her distance and possibly her sin of omission or her sin of passive participation. Now she has an idea for her paper and can read the essay documenting the techniques Orwell used to reveal distance.

These students and others should draft their papers out loud, listening to what their voice is saying, using all their experience with speaking and listening to language, tuning that voice during revision toward a piece of writing that will be heard by readers.

Voice is so central that it has its own chapter, nine. You don't have to wait until you get to Chapter Nine to read it. Browse through it now if you want to learn more about voice.

What Happens If I Discover Nothing in My Discovery Draft?

You have discovered something: that the discovery draft didn't work for you *this time*. When that happens to me, I have several choices:

1. I take a break, then return to the discovery draft and read it again, almost casually, as if it had been written by a stranger. When my mind wanders off in an interesting way, I note the wandering that may become my topic and go back to what caused it.

2. You can do a new discovery draft. Of course it doesn't work every time. But take another run at it, see where the flow of language carries you this time.

3. Try other techniques such as brainstorming to see if they work. It is hard for some students to write a discovery draft. They can't let themselves go and that's all right. They have a strong sense of form or language that inhibits discovery by free-flow writing. Fine. Not everyone of us, thank goodness, thinks the same way or writes the same way.

Does My Writing Have to Be So Personal?

Of course not, but all effective writing is autobiographical. This does not mean that everyone writes intimate confessions. It does mean that whatever you write about—football, history, computer sciences, child care, hotel management, ethics, biology—is informed by the experience you have had with the topic, no matter how academic and detached the topic may seem. We write what we know or, as Grace Paley says, "What we don't know about what we know." It is all rooted in personal experience.

Academic scholars have been studying so long that they can write personal experience papers about their researches. Of course, they do not look like personal experience papers. They are written in the language and conventions of their discipline in forms appropriate to publication in specialist journals. It is the task of each major to train students in the discourse rules of these disciplines, but the skill of writing is usually taught, at the beginning, through personal experience that can draw on what the students know from firsthand observation, discussion at home and in the street, reading, television, listening to the radio and private reflection.

Academic writing, writing from the outside in, is just as important as personal writing, writing from the inside out, in helping the writer to understand the changing world in which we all live. Writing demonstrates the discipline of the mind; it reveals how the writer assembles evidence into meaning and then develops, by critical thinking, the significance of that meaning. And writing presents the reader with a logical, documented, trail of thought that the reader can, in turn, read critically and evaluate.

───────────── ■ *Writing Exercise* ■ ─────────────

Test what I have been telling you by taking a topic from your brainstormed list and writing as fast as you can: so fast your handwriting almost skids off the page, so fast your typing leaves out letters, adds letters, runs together like a highway pile up.

Read what you have written aloud to hear a potential voice rise from the page; read to discover those failures: those sentences and paragraphs that fail but instruct, that give you a clue of what you should be saying and how you may be able to say it.

■ HOW DO I MAKE AN INSTRUCTOR'S IDEA MY OWN? ■

I started this chapter by demonstrating how to find an idea for a personal experience paper because there is a close relationship between the words "author" and "authority." The most effective writing takes place when the writer explores a territory with which the writer is familiar.

The writer should be the authority on the subject matter, but the reality of school is that we are tested by writing exams and papers in response to a teacher's assignment. And this will continue after graduation. Our world is becoming more complex and more distant. There are Japanese factories in this country and U.S.-owned factories in China. Corporations and government agencies depend on written reports sent by mail, modem, and fax machine. And the topics of most of these reports are not initiated by the writer.

Understand the Assignment

If the assignment is written, read it carefully, marking the important points so that you know what you are expected to do and how you are expected to do it. If the assignment is oral, take notes and go over them carefully. Ask questions.

It is better to appear stupid now than later. Bad writing is often the direct product of a misunderstood assignment. Good writing on the wrong topic or in an unacceptable form is still a flunk.

Interview the Assignment

What is the central question to answer or the central problem to be solved? That central issue may be stated explicitly: "Explain the relationship between tax incentives and productivity growth in the automobile industry in the United States and Japan." Or it may be implicit: "Discuss the role of government in international trade."

What assumptions underlie the assignment? For example, if this is a research course, a historical context is required, or, the ability to perform critical analysis is being tested in this paper.

What documentation or evidence does the assignment giver expect? Does the assignment giver want statistics, firsthand observation, case histories, scholarly citations or some combination of supporting information?

What are the traditions of length, form, and style the assignment giver expects? Is the assignment giver impressed with length or brevity? Is the assignment, in part, a test to see whether you can write a literary analysis using The Modern Language Association method of citation, a sociological case history, a chemistry laboratory report?

Put Your Paper in the Context of the Course

Each course, corporation, or government agency has its own environment. You should never forget that in writing you are either an apprentice or a practicing historian, psychologist, environmental planner, biologist, business manager. Each discipline has its own climate, its own expectations in written material.

A writer, to be effective, needs to know the limitations of the assignment and then discover how to be creative *within* those limitations. Every art—the business letter, the poem, the research grant—is created from the tension between freedom and discipline.

Connect the Assignment with What You Know

The effective assignment writer has performed the reading and the research demanded by the assignment, taking notes that can be read and can have sources revealed. Then the creative assignment writer has to think: to find a meaning, a pattern, a significance in the information the writer has collected.

Often the best way to do this is to connect the topic with your own experience. You may want to directly describe your personal experience with divorce in documenting a paper in sociology. More often, however, your personal experience will lead you toward research questions that need answers, and you will use your autobiographical backstage in developing a paper that is written with the objective distance appropriate to the assignment.

Your perceptions, observations, and conclusions will be influenced by your personal experience—and should be. Even a paper on tax incentives might show how your father's small-town insurance agency shifted to computers and therefore contributed to the national economy because they *are* receiving small business tax incentives for making the change. He profited, and others profited as well, because of a federal tax law. And the more specific he will allow you to be about what he spent and what he saved in taxes over the next five years, the more credibility your opinion will have with the reader. You will write with more authority if you make intellectual use of experiences that may have been emotional when they occurred.

──────────── ■ *Writing Exercise* ■ ────────────

Take an assignment you have been given in another course and interview it to discover how you can put it in a context and then connect with a subject on which you are an authority.

■ HOW DO I GET THE WRITING DONE? ■

What is the difference between writers and would-be writers?

Writers write.

It *is* that simple.

Writers suffer the same problems in getting started as nonwriters. They do not know enough; who are they to write about a subject; writing will reveal how little they know and how badly they write; they don't have the time, the place, the proper writing tools; other responsibilities have to be faced first; and on and on. All legitimate—but *writers write.*

Each writer has to develop a writing pattern and that pattern will change according to the writer's experience with a writing task, the writer's thinking and working style, as well as external conditions. But writers who write develop a discipline.

My present discipline—that I keep on a plastic card beside my computer—is:

WRITE EVERY MORNING; NEVER WRITE AFTER TWELVE NOON.

The fact I write in the morning is not important. The fact that I write at the same time every day is as important as the fact that I write when I have the most energy. Young writers may write at night as I used to but most writers become morning writers. Habit is essential to productive writing.

Writing at the same time for a few hours every day has increased my productivity. I used to have long, erratic bursts of activity, then nothing as I collapsed or dealt with all the other things I had to do that I let go during my creative explosion. A few hours of scheduled writing each day and the writing accumulates.

KNOW NEXT MORNING'S WRITING TASK.

My writing day starts when I *finish* the morning's drafting. I know tomorrow's writing task and I assign my subconscious as well as my conscious to think about it the rest of the day and night. When I arrive at my desk in the morning I find I have rehearsed what I may write.

DO IT.

Writing, like jogging or any other kind of exercise, depends on doing it. Writing is inspired by writing.

Make your own list of rules. Mine change with experience, the writing task, my living and working conditions; but discipline is essential to the writing and re-writing process.

Attitudes of a Productive Writer

The assumptions and state of mind we bring to a task often determine how well we do. If we think we will fail, we will fail; if we think the audience will laugh at us, they will; if we believe we will drop the tray, it will slip from our hands. Attitude predicts performance for writers as well as ballet dancers.

Writing Is Thinking

I write to discover what I have to say. I do not write what I have "thunk." I think by writing. I surprise myself. I learn what I know by writing.

Perfect Is the Enemy of Good

I have that sign on a shelf where I can see it as I write. I have to fight perfection—premature standards that are impossible to achieve.

A Writer's Habit

Writers make writing a habit, something they do whenever possible at the same time and place. The writing becomes expected in the way you are expected to wait on tables, show up for your job in the emergency room, deliver papers. Roger Simon of the *Baltimore Sun* explained, "There is no such thing as writer's block. My father drove a truck for 40 years. And never once did he wake up in the morning and say: 'I have truck driver's block today. I am not going to work.'"

A Writer's Deadlines

Without deadlines I would not write. I've tried it. The trick is to establish reasonable deadlines that give you a sense of achievement (I finished the draft of a novel by writing at least three hundred words a day [most days]) and forgiving yourself when you miss a day.

To do this you usually have to break a project down into parts that can be accomplished in the writing time you have. You work back from the deadline, figure the time you have, and define tasks that can be completed in a given period, perhaps the three hours you have before lunch after an eight o'clock class Monday, Wednesday and Friday. You may have six weeks or eighteen three-hour writing periods.

A two-week assignment on a paper in a literature class might look like this:

Task	Daily Goals
Read novel and make notes	Monday
	Tuesday
	Wednesday
Find, read, make notes on critical studies	Thursday
	Friday
Day off	Saturday
Scan novel in light of critical studies	Sunday
	Monday
Choose approach for paper	Tuesday
Outline	
Draft paper	Wednesday
Revise	Thursday
Edit and turn in	Friday

A Writer's Hours

Most writers, myself included, were late-at-night writers in college. This was true for me also when I first worked on a newspaper. It is a romantic time to write and I still remember the special feeling of loneliness, of being awake when others sleep, that seemed to encourage great thoughts.

But . . .

Most writers, by the time they are thirty, follow the advice of Goethe who said, "Use the day before the day. Early morning hours have gold in their mouth." Many writers have spoken of the importance of writing the first thing in the morning. John McPhee used to loop his bathrobe belt through a chair and tie himself in until the writing was done, and Jessamyn West stayed in her nightclothes until the writing was done. John Hersey said, "To be a writer is to sit down at one's desk in the chill portion of every day, and to write."

The important thing is to develop a time habit, a time when you are at your best—and to stick to it.

A Writer's Place

The importance of having a place where you are alone with your thoughts and your own blank page cannot be overemphasized. It is important to the ritual of writing to have a place where you go to work, where your papers can be left out, where you are surrounded by the tools and good luck charms that seem to encourage good work.

But I have also trained myself to work wherever I am. My notebook is always in the case I carry with me and I have a laptop computer I use as a notebook in the living room while reading, listening to music or watching television. It travels wherever I go and I can create a working environment in a motel room, in a car, on a ship, outdoors with my laptop.

A Writer's Tools

To write, you need a brain, pen or pencil, paper, and a computer. I am never without a pen, except in bed, and some small cards to make notes. I carry a bag with me wherever I go. It goes upstairs when I leave my office, downstairs when I return, and into the car when I leave the house, even if I go to the store that has my daybook and often my laptop in it.

─────────────── ■*Writing Exercise*■ ───────────────

List the conditions and attitudes that have produced your best work in school and out. Then list the problems you have writing and study how they might be solved if they fit into the way in which you have worked best in the past. Write a memo to yourself as if you were your own employee establishing a work pattern that may be productive. Post it where you will see it every day.

───

Try a writing habit on and you may share the experience of the writer who goes to the writing desk, eager for the surprise a draft may reveal when you read what you did not expect to write.

To re-write, to have a draft worthy of revision, the writer has to write; to write, the writer has to be willing to fail. No one, no matter how experienced, writes perfectly the first time. Good writing, like good science, is the product of failed experiments. Behind the best writing you read, lies a trail of instructive failures that revealed, draft-by-draft, the final text.

To write—and to do anything else that is worthwhile in life—you have to practice. My first kiss was a failure. I kissed the nose, but with practice I'm finding the target and getting better all the time. You never learn to do the important things in life; you always keep learning. Each draft, each failure, brings me its own surprise, and with a series of drafts I learn what I have to say and how to say them. The next chapter will tell you how to read so that you can discover instruction in failure.

These last three paragraphs were written *after* the final draft, on command from my editor, Laurie Runion. I felt the familiar emptiness, the fumbling along, line-by-line, and eventually discovered that I needed to echo back to the beginning of the chapter—always a good idea—and to point out that you will never learn to write and what a blessing that is. As you live your life you will discover what it means, if you relive your life by re-writing.

CHAPTER 2
READ TO RE-WRITE

To write well, you have to read well. But the craft of reading your own unfinished—often, almost unbegun—drafts is a far different process than reading the final, published drafts of writers. Their *published* writing is neat, clearly printed, polished writing. Yours is hardly writing at all in the beginning: notes, scraps of information, primitive outlines, false starts, rough drafts that begin strong, stagger and collapse. Then, there are further drafts in which meaning begins to achieve order and clarity.

How Most Beginning Writers Read Their First Drafts

Beginning writers, forced to read what they have produced, usually suffer excessive despair or pride. Commanded to premature neatness by parents, teachers and editors, they confront a mess. It looks like a failure—and it may be—but all successful writing looks like a mess in the beginning.

Reading with Despair

Writers, beginning and experienced, share that same first reaction most of the time: this is a disaster. The writer's intentions created a vision, but the reality is far different from the dream. The writer has written what the writer didn't intend to write.

The writer has to remember that the unintended writing is not failure but the starting point for effective writing. All of us who write have to keep reminding ourselves of this. Creative work thrives on surprise; but surprise, even to the creative person, usually appears as failure: the writer did not fulfill the writer's intent.

Reading with Pride

There are times, however, when both beginning and experienced writers read a draft and see exactly what they expect to see: first draft is final draft. The experienced writer may be suspicious and change just to change, feeling guilty that the writing came too easily. Writing *should* come easily and the writing that does is usually a subject the writer has rehearsed for a long time, telling the story over and over again to himself or herself, perhaps to family and friends. It is all right to read an early draft and feel pride, accomplishment, success. Revision is not a virtue but an activity that is usually necessary to make meaning clear.

Most of the time, however, writing can be improved. I can read with pride, saying to myself that this is pretty good, but then I read one word and hear a better word in my head. I change the word. I see a place where I can cut. I cut. I see where it may help to move some words around. I move some words around. I am into the draft and revising, not because the draft failed but because a good draft can be made better.

And when do I stop reading and revising? When I am on deadline.

■ READING BEFORE WRITING ■

Reading, for the writer, begins long before there is a completed draft to read. The skill of pre-reading is essential to effective writing.

Reading Your World

Writers "read" their world every day, noticing what is new, what connects, what is unexpected. And writers read their reaction to the world around; they not only notice the world but take into account their feelings and their thoughts and what they experience.

I eat breakfast with some friends at the Bagelry every morning and this morning a world famous scientist, Dr. Stacia Sower, stopped by to chat on the way to her laboratory. She is an expert on the lamprey, an unusually ugly fish that she mentioned was 550,000,000 years old, living long before the dinosaur. I was impressed and a bit depressed, feeling that the lamprey will be here long after we are gone. When I came home I started reading my feelings by writing what quickly became a poem.

An academic paper might celebrate the age of the survival skills of the lamprey but, of course, the information would be scientific, objective and documented with attribution to specific, authoritative sources. The form in which we write depends on purpose and audience. No one form is better than another. An academic report on the history of the lamprey, a research paper, a poem are all measured by their success in accomplishing their purpose for individual readers.

Reading What You Say to Yourself

I talk to myself. All of us talk to ourselves, but writers listen carefully to what we say, playing with images, details, statistics, quotations caught by words and connected by fragments of language. I watch a young neighbor learn to ride a two-wheeler and hear myself saying:

"never ridden as fast as coming down Grand View Avenue on a bike"

"learning to lean on the wind to corner"

"riding no hands"

"Slow's harder than fast."

"We have some students and faculty from mainland China. They ride slowly. The bike is transportation and they learned to ride long distances to school, to market, or to work and back each day. Americans ride for fun."

As I read what I am saying to myself, the fragments of language give me clues as to what I may write.

Reading What Others Say

Writers not only listen to themselves but also to what others say, as I listened to Dr. Sower and found a poem. Writers listen to their teachers, editors, readers before developing a draft. It is important to know what an editor or teacher expects, as well as what readers expect, but there may be a conflict here.

As we have said in the section on making an instructor's idea your own (pages 12-15), writers must often find their way *within* the limitations of the assignment or the genre. The good writer accepts the assignment: write a five-page paper on the book you have chosen for outside reading. The paper should fulfill the expectation of the instructor for topic, length, documentation and attribution, and surprise the instructor with an insight, a new observation, and unexpected question or meaning.

Reading Reading

Writers should read the best writing in the areas in which they are writing, observing the most respected writers in business, psychology, environmental science, criminology, foreign affairs, hotel management, political science, whatever, to see a range of writing appropriate to the field. Writers have to learn not to imitate—to learn from other writers without copying them.

Reading Notes

Writers make mental notes and written notes. What I read in my notes—mental and written—most often are specific, revealing details that make me think or feel. They create connections that head toward meaning—connections that may build a structure of meaning, turns of phrase that may reveal a voice.

——————————— ■ *Writing Exercise* ■ ———————————

Go to a quiet place—a corner in the library, a bench in the woods, a table at a deli on an off hour, in the car—and empty your mind. Be quiet and you will begin to see what you have passed, thinking you were not seeing, what you have overheard, what you have imagined, what has made you angry-happy-sad-content-envious-concerned-amused-worried-interested-depressed-joyful and start making mental and written notes that reveal how you have been reading your world.

■ READING WHAT IS—AND ISN'T—WRITTEN ■

When you read published writing, you concentrate on the words that appear on the page, but writers must practice a double vision. They must read what is on the page and what is *not yet* on the page.

Diagnosis Not Despair

Writers should read their first drafts with diagnosis instead of despair. One way writers do this is to ask two questions:

What Works? What Needs Work?

The order of the questions is important. Effective revision begins not with error but accomplishment, not with weakness but with strength. The writer extends what works and then deals from a position of strength with what needs work.

What works is usually found in fragments of language, floating clauses, words hidden in a tangled sentence, ideas buried under a paragraph, or a bit of language taking an unexpected flight of clear meaning before crashing into incoherence.

In each reading during revision, we read not so much for what we intended to write but for what we did *not* intend to write. The act of writing is the act of thinking, and if we are lucky our page will move beyond our intentions. When I read my drafts, I read for surprise.

Sometimes I am aware of the surprise when I am writing the draft and follow it, developing its possibilities as I write. My later reading allows me to stand back and consider the surprise: Is this what I want to say?

Other times the surprise may be hidden in a turn of voice, an unexpected word or phrase, a trail of evidence I am not aware of as I write. These surprises must be spotted and their implications considered: Is this the road I want to travel in the next draft?

The mistake is to see the surprises as mistakes. The unexpected—what you said as different from what you planned to say—is not an error. It is what happens when we write. The act of writing reveals possibilities of thought and presentation. If I knew what I was going to say in advance of writing, I would not bother to write and re-write.

An Effective Sequence of Readings

It is wise to do several quick readings of a draft, focussing on one form of exploration at a time, rather than trying to do them simultaneously.

Reading the Whole

The writer needs to step back from the word-by-word, phrase-by-phrase, sentence-by-sentence concentration essential to produce a draft and take an aerial view of the entire draft, not worrying—during this reading—about spelling, mechanics, typography, neatness. I have trained myself to become the detached reader—changing from the possessive, defensive writer of the text to a stranger who is reading what actually appears on the page—in a matter of minutes, say after a mug of coffee. But if I cannot achieve the distance I need, that is, the reader's view of the text, I imagine I am someone I know and respect who is not interested in the subject. That gives me the distance I need.

Reading the Parts

After reading the whole, I move a bit closer to see the parts of the whole, the sections that develop each part of the overall meaning. I read to find their relationship: Are they in the order the reader needs to come to the final meaning? Are they paced so the reader will continue to read—fast enough to keep the reader awake and interested, slow enough so that it is not a blur but gives the reader time to absorb the meaning of the draft? Do the proportions of the draft support the meaning—long enough but not too long? What is the relationship between the sections; do they interact in an effective way?

Reading the Line

Once the writer has the vision of the whole draft and sees how the sections fit the vision of the whole, the writer can read the draft closely, line-by-line. Each word, each phrase, each sentence, each paragraph is scanned to see how it fulfills the vision. The reading writer needs to keep moving in close but never to allow the eye to focus entirely on details. Reading line-by-line means seeing how the details support and develop the meaning the entire draft is designed to communicate to a reader. Many people read too closely too early and get lost in the details of writing—word choice, spelling, grammar, punctuation—before they have solved the larger problems of meaning, development, organization, proportion.

Reading Out Loud

And all through this, I hear the voice of the draft. Voice, the writer's word for style, is thought of as a final, superficial concern, the quick dorm room pick-up before a parent's arrival. But voice is not superficial; indeed it may be the most important element in writing. It is the way, more times than not, that I discover meaning in reading my notes and early drafts. I find that meaning is often revealed through the music of the draft the same way that the meaning of the movie scene is revealed by the musical score. The voice of the draft is tuned to the meaning of what is being written.

What *Is* on the Page

To find out what the writer, lost in the task of making, has put on the page, the writer has to perform specific reading tasks. They are:

Read Specifics

The experienced writer seeks specific details in a discovery draft that have the potential to reveal meaning, document meaning, or communicate meaning in a lively, authoritative manner.

Read this discovery draft with me and notice how the fragments of language I have put in bold stimulate my senses, my memories, my thoughts. It is the concrete detail, the authoritative face, the accurate quotation that inspires further exploration in future drafts.

> High school. Each time I drive by North Quincy High I think of the loneliness, how much I wanted to fit in, how funny looking I felt. I see the cafeteria. I loved school lunch, especially American Chop Suey that was more exotic to me than any food I got in a Scottish, New England home. Mostly I remember turning away from the serving line and without appearing to search, searching desperately for a table where I would be accepted, where I could tell **the seeing-eye dog joke**, where my **maroon sweater** would fit in, where the others would look up and smile a welcome, not turn towards each other shutting me out, where **Sheila of the red hair** with whom I **played doctor under the porch when I was nine**, and who went out with football players would see I was not a jerk but belonged in the world of men.

Read the Line

Writers must make connections between revealing specifics. The novelist E. M. Forster said, "only connect." That counsel is always good for the writer who looks for the relationships between bits of information in an early draft. A revealing specific is one that connects with other specifics. "He cheated on his income tax but didn't expect his son to cheat on his biology exam." The two specifics in this quote reveal the father.

I might find in a rough draft *"Went to New York with Mother,"* *"Dad took me to ball games,"* *"Sunday breakfasts with Dad at diner,"* *"Looking at late movies with Mother,"* and I see a pattern of distance I wasn't aware of as a child. That might lead to a personal piece on why I didn't see what was happening, why I wasn't surprised at the divorce, why they grew apart, or a sociology and psychology paper on children's lack of awareness of stress between their parents.

Read Fragments

Often potential meanings are hidden in a word, a phrase, a line. I have had to teach myself to read fragments, and so should you. When we read our drafts we are like the archaeologist who finds a fragment of a bowl, the preserved ashes of a fire, a sharpened piece of stone, and then uses a trained imagination to create a civilization.

There are many forms of fragments that reveal meaning. Here are some that I frequently discover:

Code Words. These are words that have a private meaning for us. I read "basement," an ordinary word, and suddenly remember that was the word in first grade that took on a new meaning and a terrifying, seductive mystery. When I went to first grade the basement was where you were sent when you raised you hand and said, "I have to go," and it was where other little boys told you . . . .

Revealing Detail. When I had my heart by-pass, I was part of a machine for 91 minutes. That we can survive while surgeons diddle around with such a vital organ as a heart opens the door to the poems and columns I have written about that experience. It could lead, in my hands, to a novel; in another writer's hands to a screen play, a non-fiction book on heart surgery, a play, a biography of a surgeon or an autobiography of the patient.

Significant Phrase. In New England, we all look forward to "Indian Summer," those days in the late fall when the weather suddenly turns warm. But the phrase was born in fear. In colonial times the pioneer farmers worked their fields with musket nearby and huddled at night in crowded forts where they could protect their families against attack by Indians. When it started to get cold, the Indians would not attack and the families could live in their homes beside their fields. Then the weather would change and the Indians would return, surprising individual families. That phrase has for me two hidden meanings, meanings that have a tension filled with possibility for writing.

Haunting Image. After we buried my father, I went into my parents' bedroom and found my mother sitting alone on the edge of their double bed that had been worn to the shape of their double forms. I am haunted by that image. We all have those snapshots of memory that carry enormous loads of meaning we may choose to explore and share by writing.

Read Other Clues to Meaning

There are many other elements than fragments that reveal meaning to the writer. Some of the most significant to me are:

Music. Writing is heard. Writing is speech written down, and writing is *heard* more than *seen* by the writer. The melody of the writing often tells me where the meaning is and what it is. I hear my own words telling me that an event has more significance than I realized, and this evolving meaning makes me angry, sad, happy, amused, instructed.

Pattern. Scanning a draft, I see a pattern of possible meaning emerge. Information gathers in clusters, divides along an unexpected fault line, engages in civil war with other information. I see in the draft a direction, a pointing towards meaning.

Thread. When I read a draft fast, I often begin to see a thread woven through the text that I was not aware of while writing. Images, phrases, metaphors, echoing details and words lead me toward a meaning I did not expect, but am compelled to pursue. I am taught what and how to write by the thread that runs through the draft.

Read to Find a Problem

Humans are problem-solving animals and we solve many problems through that most disciplined form of thinking we call writing. The draft connects details so the writer, then the reader, discovers a pattern of meaning.

Read to Discover What *Is* in the Draft

It is easy to read with despair, but the writer must remember to read to see what is in the draft that works or is filled with potential. It may be the voice, the organization, the information, the point of view—any of the many elements of effective writing. Remember that we revise most effectively by developing what works rather than by correcting error.

Read to Discover What Is *Not* in the Draft

I think it is important to read positively, to recognize the potential within the draft, but there are drafts in which no potential is apparent. It is neither on the page nor in the world from which the draft was written. Beginning writers must read to see what is not even in the rough draft.

Drafts *without apparent potential* usually display some or all of these deficits:

Problem: No Territory

The writer has not created a world of people, events, or ideas the reader and the writer need or want to explore together. An inviting world has a richness and a complexity that interest the reader.

Solution: Move the location of your article to a place where you are familiar enough to know the simplicities and the complications that interact to make good writing. A paper on ethics might be moved from the Supreme Court, where you haven't served, to the football field, where the coach has just asked you to cheat by faking an injury. On the other hand, it is equally possible to write about faking an injury in the context of National Collegiate Athletic Association rules, emphasizing research rather than personal experience.

Before: Often the United States Supreme Court has to rule on ethical as well as legal issues.

After: I am only able to go to college because I have a football scholarship but I'm not a star. I'm special teams on kick-offs and punt returns; I hold the ball for the place-kickers; I carry the ball in certain short yardage situations; and, although we are a religious school, I am the quarterback's retaliator and the guy who fakes an injury on the coach's command.

Problem: No Surprise

The reader reads only what the reader expects. The reader knows what is coming next. There is no challenge, nothing that provokes a thoughtful or an emotional reaction from the reader. There is no suspense.

Solution: Look over your notes, written and mental, to see what surprises you, what you learn, what you question, what your reader needs to know, what runs against expectation, to find a subject that will interest a reader.

> **Before:**   Research into the nation's early history may change the contextual environment of our mythic beliefs.
>
> **After:**   We think of "Indian Summer" as a delightful return to summer in the autumn before winter sets in, but the term had a different context in colonial times. Then settlers feared the unexpected warmth because Indians, who did not attack in cold weather, suddenly returned to raid isolated farms.

Problem: No Writer

The reader does not sense the presence of a human being behind the draft. The draft does not breathe; there is no individuality. There is none of the essential human music we call voice that is essential to the individual act of one person, a writer, meeting another person, a reader, on the page.

Solution: Read the draft aloud and tune it until you find a voice that sounds like you and *is appropriate to the meaning* of what you have to say. This may mean writing in the first person—I—or in the third person—he or she. Whoever the speaker, the reader needs to sense a single human intelligence behind the words on the page.

> **Before:**   One is rather overwhelmed when one arrives on a university campus from a small community for the first time.

After: I chose the university because I came from Eagle Pass, population 105, and was graduated second in a class of seventeen in a county high school. I wanted to escape. I knew everyone and everyone knew me. Watch out when your dream comes true. Here I know no-one and no-one knows me. I'm doing solitary.

Problem: No Respect

The writer does not respect the subject matter, the reader, or him- or herself. The writing task is not taken seriously, but kissed off, dependent on superficial tricks that call attention to themselves, not to the subject. The writing cheats the reader. It is dishonest.

Solution: You are revealed when you do not take the subject seriously. Find a topic that you can take seriously and write with the heart as well as the head. Look for a way to respect your topic, to have compassion for the people involved, to respect opinions you may not share.

Before: In an election year, politicians talk of entitlements but nothing is ever done to cut off these pay-offs to free loaders.

After: I know there is welfare fraud, but last week-end we put my grandfather in a nursing home. He worked hard all his life, voted Republican, even refused his veteran's bonus after his war, but the care he needs costs $3,500 a week, without extras. He worked as a truck driver. Now his bill will be $182,000. He needs Medicare, Medicaid, Social Security, anything he can get.

Problem: Too Little

The reader is given nothing but generalities. There is no information, just emotional or intellectual generalities that pass over the surface of the subject. There is no evidence that allows me to think *with* the writer. There are no revealing, resonating details.

Solution: We often feel that writing is most intellectual when it is full of generalities, but theories must arise from fact. Readers are hungry for an abundance of accurate, specific information that allows them to do their own thinking. Here the writer uses a personal anecdote, but the same information could be delivered in statistical terms by the writer's investigating bank records or government reports.

Before: Our economy is fueled by bank loans but sometimes there is inadequate investigation by loan officers.

After: My father runs a tractor repair and sales business and he needs seasonal bank loans to tide him over until his customers sell their crops. We need a bank that knows us, but I was shocked when a local bank got into trouble and it came out that almost 70 % of the loans were to relatives of bank officers—or to the officers themselves. That's a bit too local.

Problem: Too Much

The writer floods the reader with too much information of equal importance, or pours so much rhetoric on the page that the reader drowns in language.

Solution: Whether you are writing a term paper, an essay, or a story, you should be careful not to load up on so much specific information that the hearing of each piece is lost and your writing becomes a jumble of unrelated information. You should decide on your single, dominant message and then cut anything that doesn't move it toward the reader.

Before: The party started early, before the football game, at lunch, with beer and then the guys got booze into the game in hip flasks and then we all went out to the lake where Flynns serves the best steaks but it was crowded and we had to wait an hour and a half in the bar, where, of course, there were girls and they'd had a few. We did eat lots peanuts and I saw my priest there but I don't think he saw me. He was pretty red-faced. Well, that's not the point. The point is that Rafe outdrank us all. He always did and there was no talking him out of driving, never is. Two dead.

After: Saturday I lived a TV commercial against drinking and driving. We would never do that, but we did. We drank from 11:30 in the morning until 10:45 at night when the police, the wreckers and the ambulances arrived at the intersection. Two stretchers left with the sheet over the faces of two of my best friends.

Problem: Too Private

The writer is so close to the subject that the reader has no idea what the writer is talking about. The writer produces a mumbling monologue of code words and phrases that may mean something to the writer but are obscure to the reader. Private writing is a particular problem when the writer becomes, at least in the writer's own mind, an expert on a subject, such as Internet, honey bees, systems engineering, the history of the Dred Scott case.

Solution: Put your writing in context. This can be done with a short paragraph, sometimes with a sentence, sometimes with a phrase. In writing about an individual experience with rape, you can remind the reader, with statistics, how common this crime is on campus.

Before: When I booted on my 486 and clicked graphics mode in WP 6.0 I entered a WYSIWYG environment.

After: When I turned on my computer and selected WordPerfect's® most recent word processing program, I was able to move to the graphics mode and see underlining, bold print, italics and different type faces as they will appear on the printed page, a feature not available in earlier versions of the program.

Problem: Too Public

The reader is embarrassed by the writer who tells readers more than they want to know about the subject, often providing an excess of inappropriate, intimate details that do not seem to relate to the topic.

Solution: It may help to write about the subject in the third person, combining personal documentation with more objective sources, standing back just a bit from the subject.

Before: My husband wanted sex before breakfast every day. He'd been so sweet before we were married and I like spontaneity, but not the same spontaneity every day, after jogging in the afternoon, Friday and Saturday night after parties, especially when we visited home and my parents were in the next room. Him on top, me on top, in the shower, in the car, all the strange stuff. He had a book and it was like were taking a course, chapter-by-chapter.

After: She didn't think her marriage could be saved because her husband who had been her colleague at work, saw her only as a sex servant after marriage, but Couple Therapy helped him see how she felt and they worked it out.

Problem: No Significance

The information the writer provides is not put in perspective. There is no emphasis, no clear point. The writer is just delivering information. There is no evidence the writer has thought critically about the subject and what it means.

Solution: Writing is always a form of critical thinking. The reader expects more than "just the facts, M'am." The details must add up to something that affects and involves readers, making them think or feel—or both.

Before: A foxhole in infantry combat holds just one person. Close together, makes you vulnerable to mortars, hand grenades, mines. Extended order means you march with several yards between you. Taking cover from each other, you disappear from your own army. Greater fire power means a single soldier can command a large field of fire.

After: Infantry combat is lonely. In the movies, soldiers huddle together where the camera can see them interacting. In combat, the soldier is alone in the foxhole, dug far enough apart that one shell hits only one foxhole. Each soldier's first enemy is loneliness.

Problem: No Connection

The draft is neither placed in a larger context—political, historic, sociological, psychological, scientific—nor does it connect with the experience of the reader.

Solution: References can be woven within a draft to connect with the reader's experience, using business examples and references for a readership of business people, sports references if the readers follow sports.

Before: Poverty is terrible. Bad housing crowded together. People sitting an afternoon away on doorsteps or staring out the window. Kids playing without toys. Hopelessness.

After: Studies by Murray (1989) and Morison (1991) have demonstrated that television has brought the poor into intimate contact with the affluent, so that the "have nots" are forced to see how much the "haves" possess—or are invited to purchase—many times in each hour. Starobin (1992) has confirmed what Nestelberger (1980) theorized: that the poorer the home, the more hours the television is likely to be turned on.

──────────── ▪ *Writing Exercise* ▪ ────────────

Take a draft, perhaps on a subject you are drawn to but haven't been able to write about successfully, and read it to see the potential of what is and what is not on the page. Read through this last section, doing each reading I suggest. Of course, you will not read a draft in this formal, military fashion. But try it once to see what it reveals that may be developed as you re-write this draft in the future.

▪ READING A FINAL DRAFT ▪

Earlier readings should be quick, but the final reading must be slow. I find that I have to break this final reading into fifteen-minute units of intense concentration.

Reading for Communication

Readers come to a draft with expectations. They expect words to be spelled so they can recognize them; they expect the pattern of the sentences and paragraphs to be familiar; they expect punctuation to make those patterns clear; they expect each word to serve a role in revealing, developing, defining and clarifying meaning.

Grammar and mechanics are not in the fashion business, but in the meaning business. As Ernest Hemingway said, "Prose is architecture not interior decoration." The traditions of language represent what we—writer and reader—have learned about communicating meaning. The conventions of language are designed to make communication easy.

When a draft is final, the writer reads word-by-word, line-by-line, to make sure that every word, every punctuation mark, every space between words, clarifies meaning. This reading is so intense that I often have to break it into ten- or fifteen-minute chunks so that I attend to what is on the page, not to what I hope is on the page.

But line-by-line reading is not a chore for me, but fun. It is satisfying, as meaning comes clear under my hand.

Reading for Flow

After I have read the final draft line-by-line, I read it through again for flow. Effective writing should carry the reader along as a person in an inner tube is carried by the flow of a river, with ease and at an interesting pace. Flow is composed of the writer's ability to:

- anticipate and answer the reader's questions as they are asked
- adjust the pace so the reader moves through simple material fast enough to avoid being bored, yet slow enough to allow understanding of difficult passages
- tune the music in the draft's voice so it supports the evolving meaning of the draft
- provide information that satisfies the reader's hunger and makes the reader want more
- run a clear narrative line through the draft that will carry the reader to meaning.

CASE HISTORY
Reading a Professional's Draft

■ ════════ ■

Recently the sports editor of *The Boston Globe* asked me to do an opening-day piece on my first visit to Fenway Park to see the Red Sox play. It is always fun to do a nostalgic piece and I agreed immediately.

I had the instigating line which I discussed in the last chapter. It was the very Fenway Park that I had first gone to when I was about six years old, and which had been the location of so many night dreams and day dreams and staring-out-the-window school-room dreams.

I also knew I would make the assignment my own because the editor's direction was to write a personal piece. The only limitation was that it should be written about my first visit from the perspective of my present age. That gave me plenty of room. If the editor had invited a historical piece I would have looked up newspaper accounts of the game or even gone back to the days before Fenway Park, when an uncle of mine used to stand behind a rope—there were no stands—to see major league baseball. I could have written about the historical changes in the game. I could even have written about the architecture of baseball stadiums and how the rules of the game had to accommodate the odd shapes of baseball stadiums built on odd-shaped city lots.

I knew the instigating line would take me somewhere but I didn't know where. When I got up the next morning to explore the line, I wrote a discovery draft and I found I was not so much writing of the Red Sox but of my father:

> When I think back to my first visit to Fenway Park, and all the visits since, I think of my father, who had no childhood, and how the Red Sox was the one constant thread of communication in our lives.

I heard a voice in the last part of the sentence:

> "my father, who had no childhood, and how the Red Sox was the one constant thread of communication in our lives."

This was rich material—*"a father who had no childhood"*—and my childhood, in which I often felt I had no father. There was tenderness and understanding and anger and regret all in the way I heard those words when I read them aloud. I had, of course, to ask myself whether I wanted to be so personal. My father was dead but would he want me writing of us and of Fenway Park? I thought he would and I went on.

It would have been easy for me to read these first lines with despair. It didn't sound like the usual first visit to Fenway Park that the assignment dictated. It didn't focus on baseball but on my father. However, I have learned to accept, even cultivate this sort of failure. I had two territories to explore: the father who had no childhood, and the relationship of father and son.

I knew I had been reading my world, my relationship with my father and our shared passion for the Red Sox. I included material on Fenway Park and who played for the Red Sox when I was young, but the real explorations were in family, not sports, memories. I saw my father and our relationship differently than I ever did before as the first draft unrolled. As I read what was on the page and what was not on the page I found myself read—as I wrote:

> My father was always at the store, off to New York on business or at Tremont Temple for morning and evening services, for Wednesday night prayer meeting and for church business meetings in between. He left home before breakfast and came home late. My uncles were more father to me than this quiet, mustached man with thick glasses and a sad smile.

And later I read in my draft our childhood relationship, as complicated as most such relationships are, in a way I had not seen it before:

> I knew then, in some strangely reversed way, that I was my father's instructor in childhood. He was the oldest son of Scots who indentured themselves to Fall River mills to get to America, and I could never get him to remember a single childhood game.
>
> Stern economic facts and equally harsh Calvinist beliefs made life a serious business. He had wanted to be a teacher or a minister, but his father, who was to die young and pass his paternal responsibilities

on to his eldest son, took my father out of school on the morning of his fourteenth birthday in 1904, the moment he could legally be removed from the classroom.

I taught my father to toss a baseball, but he always tossed it like a girl, in a day when girls threw baseballs like a girl. Father never got the hang of a football and on each two-week summer vacation in Maine I tried to teach him to swim and failed.

As I came to the end of the draft, I saw how through all the conflict and pain of my leaving his church and his political party, even while I deserted his ambitions for me and lived a writer's, even an academic life he could not understand, that we had one sure line of communication that was always open.

Now looking back, twenty-one seasons after his death, I realize that when we could not talk politics or religion, when he could not understand the dreams of a writer and the worlds of a reporter-turned-professor, we could talk baseball. It was the one certain world we shared.

That first day at Fenway opened lines of communication that lasted all our days together. He talked the Red Sox to me when he came home late from a trip to New York and found me sick again, and he could not talk of the fear he must have felt for his only child. And I talked the Red Sox to him the time when it was my duty to say "cancer," when we had to discuss his failing heart and the time for finally letting go.

Throughout all the drafts of this piece I read what I did not expect and ended up writing a piece that was much more than the familiar story of a first visit to a ball park.

As I wrote it, I had to make sure I had taken hold of a territory. My territory was the world of sports shared by a parent and a child. I felt that through the personal I was reaching the universal, that my readers would see themselves in my experience. I hoped I showed respect for my father and not attacking him. I hoped there was enough specific information to satisfy the reader and enough distance from my father's relationship with me not to embarrass the reader.

Readers wrote me after it was published to say that my reading of what I was writing sparked them to their reading of their own experiences as fathers, mothers, sons, daughters. That is one of the great magics of writing. As we tell our stories, we liberate the stories of our readers.

Each draft is a river-rafting ride into the unknown. We have to lose control of our writing to complete a draft worthy of control. There is an essential dumbness to good writing, a need to begin in ignorance to discover what we know and need to know.

When I complete a draft and I sit back to read it, I have the usual sense of despair, of hopelessness, of not achieving what I had planned to, but then the draft sweeps me along from despair to surprise: I have something to say that is worth saying. Now I will apply the re-writing craft to continue the play—the critical thinking—that is at the center of re-writing.

CHAPTER 3
RE-WRITE TO WRITE

The myth: The writer sits down, turns on the faucet and writing pours out, clean, graceful, correct, ready for the printer.

The reality: The writer gets something—anything—down on paper, reads it, tries it again, re-reads, re-writes, again and again.

For years I denied the reality. I held to three firm beliefs:

- First draft was best. Good writing was spontaneous writing.
- Rewriting was punishment for failure. The editor or teacher who required revision was a bad reader who had no respect for my spontaneous style.
- Revision was a matter of superficial correction of error that forced my natural style to conform to an old-fashioned, inferior style.

No one challenged my literary theology; they simply didn't publish me. And I needed to get published. I wanted to eat.

Bob Johnson of the *Saturday Evening Post,* then the leading free-lance market, liked what was in an article of mine. He said they would hire a writer to fix it up.

"I'm a writer," I said confidently.

"Well," he said doubtfully.

"I want to write it myself," I pleaded. "Please," I begged.

43

He sent me a single-spaced letter of criticism that was longer than the article I had submitted. I re-wrote, and then re-wrote what I re-wrote.

He traveled from Philadelphia to Boston to go over my revision word-by-word, line-by-line. Again I re-wrote—and re-wrote.

The article was published and then I began to listen to what my editors and the publishing writers I began to meet told me: revision is a normal and essential part of the writing process. I began to pay attention—and be comforted—by the testimony of hundreds of the best writers of past and present:

> *Because the best part of all, the absolutely most delicious part, is finishing it and then doing it over. . . . I rewrite a lot, over and over again, so that it looks like I never did.* TONI MORRISON

> *I like to mess around with my stories. I'd rather tinker with a story after writing it, and then tinker some more, changing this, changing that, than have to write the story in the first place. That initial writing just seems to me the hard place I have to go to in order to go on and have fun with the story. Rewriting for me is not a chore—it's something I like to do. . . . I've done as many as twenty or thirty drafts of a story. Never less than ten or twelve drafts.* RAYMOND CARVER

> *My writing is a process of rewriting, of going back and changing and filling in.* JOAN DIDION

> *When I see a paragraph shrinking under my eyes like a strip of bacon in a skillet, I know I'm on the right track.* PETER DE VRIES

I began to hear the message: *writing is re-writing.*

■ WHY SHOULD I RE-WRITE? ■

Revision is an adventure in meaning, the exploration of content. Through revision, the writer explores the worlds of memory, observation, speculation—the worlds of events, people, ideas.

Re-Write to Find Out What You Have to Say

The main reason to re-write is to discover the full depth and dimension of what you have to say. Each morning, as I write and re-write, I discover that I know more than I thought I knew, and I learn as well what I need to learn. I surprise myself as information connects to create a new insight, a new possibility, a new meaning.

Writing is thinking, the most disciplined form of thinking. And thinking evolves from what you know to what you are beginning to know. There are thoughts—usually vague—hints, clues, "what-ifs" that precede the words on the page. They change and begin to clarify as they are captured in a first draft. The writer has an idea of what he or she *may* say.

Re-Write to Understand What You Have to Say

The writer re-writes to understand the first drafts. Usually the writer begins to understand a central or primary meaning and that causes the writer to both eliminate what does not relate to that meaning and to see new possibilities in probing into the meaning and extending its horizons.

Beginning writers and some not-so-beginning writers are often impatient. But the famous essayist E. B. White says, "Delay is natural to a writer. He is like a surfer—he bides his time, waits for the perfect wave on which to ride in." And novelist Virginia Woolf says, "As for my next book, I am going to hold myself from writing it till I have it impending in me: grown heavy in my mind like a ripe pear; pendant, gravid, asking to be cut or it will fall."

They know it takes patience to write and the best way to wait is to play with the draft, taking out and putting in, writing over, moving things around, working it until you begin to understand what the later drafts have to say.

Re-Write to Make What You Have to Say Clear to Readers

After you, the writer, know what you have to say, then you can make it clear to a reader. Often the impatient writer attempts final revisions and editing prematurely, trying to make clear to a reader what isn't yet clear to the writer. It never works. The draft simply becomes more and more confused.

Writers re-write because that is where the real excitement is. Discoveries of significant meaning are rarely made in the head. Instead, they arrive on the screen or the page during the intellectual and artistic play of revision.

■ THE RE-WRITER'S ATTITUDE ■

Our attitude often predicts the result of what we do. If we come to re-write, feeling it is punishment for failure, that nothing can make our draft better, we will not produce a series of improving drafts. The effective writer knows that revision is not the product of failure but a normal and necessary part of the writing process. The writer comes to the re-writing task with positive feelings, knowing that the act of revision will produce its own excitement and motivation: we will read what we did not expect to write.

Welcome the Unexpected

Beginning writers—and some experienced ones—forget that writing is thinking, not thought reported after the fact. Most of us are conservative. How many bosses have said, "I don't like surprises"? Yet surprise is at the center of the writing / re-writing process.

Of course as we write and re-write, think and re-think, what we have to say will change. And the most important breakthroughs in understanding come at the point of surprise. We start a paper on the California Gold Rush in 1848 and find that what fascinates us is the fact that the population of many areas of rural New England dropped and whole towns disappeared. Our paper turns from being what happened to those who went to California, to what happened to those who were left behind in New Hampshire. It is about how the loss of population affected New Hampshire agriculture, the economy, the economic and political development of the area that was deserted.

The effective writer courts and appreciates the unexpected, the unplanned, the contradictory, the surprise.

Strengthen What Works

The weakness in the first draft of the paper on the California Gold Rush might be the information about California. The strength might be the information about abandoned towns in New Hampshire. It may be necessary to get a teacher or an editor's approval to shift the topic, but that is the piece that is ready to be written. Most teachers and editors will be fascinated and approve.

The strength may be in the discovered topic. It also may be in the quality of specific information the writer has, the pattern or order of that information, the discovery of an unexpected audience, the passion or curiosity of the writer, the place in the draft where the voice is strongest.

The experienced writer knows that the first task of revision is to strengthen what is effective so it becomes more effective—and that process of strengthening may also solve many of the problems of the previous draft.

─────────── ▪ *Writing Exercise* ▪ ───────────

Write down the attitudes you have been taught and developed about re-writing before beginning the course, then write out the attitudes you think you need to be an effective re-writer. Put them over the desk where you do your revision.

▪ THE PROCESS OF RE-VISION ▪

Beginning writers try to revise backwards. That is usually what they have been taught intentionally or inadvertently by teachers who pass back papers with grammar, spelling, mechanics, typography, criticized or corrected. Those are important concerns but they come at the end, not at the beginning, of the revision process.

Revision means to see again and that involves looking at the entire draft and solving its problems in sequence. There is an order to revision, a priority to the problems that have to be solved.

The experienced writer may solve many problems simultaneously but each needs to be broken down into a logical order for the inexperienced writer to understand and practice them, the way the tennis coach breaks down a sequence of overlapping moves the player has to put together to volley.

There is a chapter on each of these stages of revision in *The Craft of Revision* but to understand their relationship to one another, it is important to scan an overview of the revision process. Here is the process defined and then briefly described: **FOCUS, COLLECT, SHAPE, ORDER, DEVELOP, VOICE, EDIT**.

The Basic Re-Writing Process

Re-Write to Focus

After reading a draft, the writer has discovered the primary meaning of the draft. That potential meaning is the **focus** essential to effective writing. It is a good idea to write that meaning down in a word, a phrase or a sentence. Everything in the draft must lead to that meaning or follow from it.

Re-Write to Collect

Writing is built from an inventory of specific details—accurate, revealing information. This information is collected from observation, memory, experience, interviews, reports, books, articles, tapes—a vast abundance of resources. The authority and the liveliness of a piece of writing depends on specifics and the art of revision consists of being able to **collect** an abundance of significant information that relates to the focus of the draft.

Re-Write to Shape

Many students have never thought of the shape of a piece of writing, certainly have never considered shaping their own material. Teachers have given them a form—for example, the five-paragraph theme, the book report, the lab report—that is similar to the wooden form into which concrete is poured. It is to be followed without question.

The writer, however, often shapes the writing form to make the evolving meaning clear. This shaping is often done by a system of outlining that will allow the writer to see the skeleton of the final draft.

Re-Write to Order

When the shape or form of the writing is established, the writer can attend to the structure within the draft. When the writer knows the message, the reader, and the traditions of the form, then the writer can select the **order** in which the information should be delivered.

Usually the writer can anticipate the questions the reader will ask and answer them within the draft. Other times a traditional form will predict the reader's questions. For example, a chemist may expect an abstract at the beginning of a scientific paper; a doctor may expect specific patient observations at the same point in each nursing report; a judge may expect legal precedents at a special place in a brief to the court.

Re-Write to Develop

The reader hungers for well developed information: the statistic that is in a meaningful context, the anecdote that humanizes and illuminates a fiscal crisis, the quotation that sums up an argument. The writer, close to the subject, often feels that the reader will understand the draft because the writer understands it.

The reader needs a fullness of information, a draft with texture. And the writer, in satisfying the reader by developing a draft, learns more about the subject than the writer expects. The process of exploring the subject continues.

Re-Write to Voice

The writer listens to the voice of the line in the writer's head before there is a line on page and at each stage of the drafting and revising, the writer listens to the voice of the text. But, just before editing—and voice is essential to effective editing—the writer revises the almost completed text by "voicing" it, making sure that the voices in the text work together in harmony to create music that will support the meaning of what is being said.

Re-Write to Edit

The process of learning continues as the writer edits the final draft so that the reader will understand what the writer has to say. While editing, the writer stands back from what the writer expected to say so the writer can see what the text actually says and must say.

Now that the greater problems of revision have been solved, the writer reads the draft line-by-line, making sure each word is the most effective word and that it is spelled correctly, that the sentences and paragraphs carry significant information to the reader in a clear and pleasing manner, that all the mechanics, such as punctuation, reinforce meaning, that nothing gets between the writer and the reader.

Look into the text to discover its strongest, most interesting parts. Revise to make their meaning clear to you—and eventually to a reader. The draft will instruct if you follow it. It will ask for a definition here; more description in one place, less in another; an increase in specific detail on one page, less on another; the elimination of good material that slows the reader down; the insertion of material that answers the reader's questions. The draft will tell you what to do if you listen to what it is saying and fool around with possible solutions, playing with words until they make your meaning clear.

CHAPTER 4
RE-WRITE TO FOCUS

When the ambulance screams up to an accident, the paramedics have to decide which patient is most seriously injured and then they have to decide what is the victim's most serious problem. There are usually cuts and bruises, broken bones, external and internal bleeding, difficulty in breathing. The medics have to decide immediately which problem threatens life and what can be done about it.

When I read first drafts from students, from the professional writers I coach, from my own computer, the priority, draft-threatening problem is my focus.

I may see a draft that is sloppily prepared, full of misspellings (and if this poor speller can catch them, the writer is in real trouble), punctuation problems, incorrect facts, illogical structure, and an uneven, awkward style, but I have to probe for the real problem. It is, most likely, a lack of focus.

Oftentimes, I see a slick, polished piece of writing that just doesn't add up. The typing is beautiful, the manner clever, with well-turned phrases; there are no errors of fact or grammar, but I don't know what it all means.

Another frequent problem is the carefully wrought draft that is totally without surprise. It is dull, tedious, predictable, boring. Nothing is wrong but, yawn, who cares? I know what the draft will say before it says it.

■ DIAGNOSIS: NO FOCUS ■

To find out if your draft has a focus, read it fast without worrying about organizational or language problems and then write the focus of what you have read, in a single sentence.

The Elements of an Effective Focus

An effective focus should be clear. You shouldn't have to think a long time after reading a draft to find a possible focus. Readers will not make that effort; they will just stop reading. Right now, all you want to do is to see if there is a focus.

To discover and state the focus, you should be able to answer the following questions with a specific, brief statement—a sentence or less.

What Is the Single, Dominant Meaning?

Every piece of effective writing will say many different things to individual readers. Good writing has depth and texture, but something should predominate.

The account of a beach party may include interesting material on who was there; how they dressed; how they behaved; what, and how much, was eaten or drunk; what the swimming was like; what games were played; who came with whom; and who went home with whom; but the account must have a meaning, and the meaning might be hidden in the abundance of all the details of the party. And these details may appear joyful, until the person looking back finds he or she is writing of a "date rape."

What Is the Central Tension Within the Dominant Meaning?

The problem of date rape is a serious one but that label will not attract, hold and influence readers. Effective writing contains a central tension that puts everything in the article in a different perspective. That central tension might be "Beer doesn't make no, yes." The alcoholic hilarity of the outing suddenly turns false and dark—ominous, not celebratory.

The argument might focus on the reader who thinks women say "no" and mean "yes" and therefore a few beers "to loosen her up" justifies a sexual attack: she really wanted it. The account might show how dress and behavior do not justify sexual assault. There is tension in the topic because the attacker sees a pattern of seduction that the victim does not intend.

What Do Test Readers Say Is the Meaning of the Draft?

It is often helpful to get someone who is not familiar or sympathetic to read a draft and then tell the reader—in a sentence—what it means. I am careful about the person I choose to be a test reader. The person does not need to be an authority on the subject and, in fact, it is often better if the person is not, as my readers will not be. My test readers are not always writers, but they are people who can listen to what I want of them, and respond helpfully to an early draft. If I ask simply if they get the meaning, then they do not immediately jump in and correct my typing and my spelling. They give a candid response to the question I ask. Most of all, the test readers I return to are those who make me want to write as soon as I leave them. They may praise or criticize, but they inspire me to go to work revising the draft.

When a test reader gives me a meaning I did not intend— "Rape is okay at a beach party"—my first tendency is to say, "bad reader" but that is not good enough. The writer has to communicate to readers who are rushed and harried, not interested in the subject, or opposed to the writer's views. It is the responsibility of the writer to create a focus that will be clear to many readers, good or bad, interested or disinterested.

Sometimes there are poor or eccentric readings of a draft, but most times when a test reader gives a meaning you do not expect, you can scan the piece and discover the focus is not what you meant. Then you can point the draft in the direction you want.

If the Diagnosis Is Positive

If the draft has a clear focus, then move on to the next step in the revision process. Each draft will have its own problems as it passes through revision. The effective writer moves over those stages in the process where a quick diagnosis reveals no problem.

■ SAY ONE THING ■

One of the biggest differences between the successful writer and the unsuccessful one is that the successful writer says one thing. One idea dominates.

The writer may have known that single idea before writing. It may have come clear to the writer during the writing. Or the idea may be discovered through the reading of the draft. But once the idea is recognized, it has to be developed and clarified by revision.

How Can I Find That One Thing?

It is vital to articulate the one thing that brings all the issues in the subject into focus. Some of the techniques I use to do this follow.

Questions to Reveal the Focus

Sometimes it is a good idea to back off, turn the draft over or store it so it is not on the screen. Then think about the subject. After writing and reading the piece, ask yourself:

- What surprised you?
- What did you expect to read? How was what you read different from your expectations?
- What do you remember most vividly?
- What did you learn from the writing and the reading?
- What one thing does a reader need to know?
- What is the single most important detail, quote, fact, idea in the draft?
- What do you itch to explore through revision?
- What single message *may* the final draft deliver?

Sharpen the Focus

Once you have an answer to those questions, then you can sharpen the focus. Some rules for sharpening the focus:

- Use as few words as possible.
- Play with specifics from both sides of the issue that is in tension; avoid generalizations and abstractions.
- Use nouns and verbs, especially active verbs.
- Reveal the central tension.

For example, a student might write: "I'm against violance of any kind. In my experience, and in the opinion of clergy of all faiths as well as sociological studies, it doesn't work. Forget if human beings get maimed or dead and all those human concerns. Figure some people got to get hurt. But I know, from personal experience, that hurting someone physically hurst them mentally but does not make them behave. And it may make the victim as well violant, continuing the circle." *After a conference with a teacher or classmate who said,* "I'm confused. What are you trying to say—what is the writing telling you to write?" *the student might try a more focussed approach:* "The more my father beat me to keep me off the streets, the more I fled to the streets. We live in a violant society and the arguments against violance are all idealistic. Mine argukmemnt against violance is pragmatic, no idealism involved: violance doesn't work."

Titles and First Lines

The focussed meaning may, in fact, become the title. When I was freelancing magazine articles, I would start writing an article by brainstorming one hundred to one hundred fifty possible titles, in fragments of time as the research was winding down. Each title was a window into the draft I might write.

To brainstorm, you have to be willing to be silly, knowing that in this freedom may lie an important insight. If I was assigned to write a paper on roommates, I might start with titles that would remind me of experiences—and problems and conflicts and satisfactions—I had with roommates.

My Roommate for Forty-three Years
My Forty Army Roommates
Why We Had a Fistfight

My Roommate's Smell

Snores

Snores and Bores

Don't Room with a Philosopher

Familiarity Breeds Familiarity

Ten Rules for My Roommate

How to Drive a Roommate Crazy

Talks ok—But at Three AM?

My Roommate's Snake

My Roommate's Brother

Why I Murdered My Roommate

Why I Murdered My Roommate—and was Acquitted

Roommate or Cellmate?

Cheese, Toothpaste and Computers

The Music Wars

Jazz Rock Folk Classical

Rocking to Mozart

One Roommate and Four Alarm Clocks

When His Lover Stays Overnight

The Importance of Privacy

No Passion Please

Living in My Roommate's Plant Jungle

The Poster Wars

Borrow My Boyfriend, Not My Jeans

What's Hers Is Hers; What's Mine Is Hers

You could go on and on and so could I. I may, for example, write a humorous column about my wife, my roommate for 43 years in the form of advice to freshmen meeting their roommates for the first time.

─────────── ■ *Writing Exercise* ■ ───────────

Write possible titles as fast as possible in specific fragments of time. I used to do 150 or so at a run. And what if number 3 of 150 is the best? Well, now you know it! The discarded titles may turn up as lines in the article or as starting places for other articles.

The fragments of language that focus meaning often become the first lines of a piece of writing. As a journalist, I am a great believer in writing the lead—the first line, the first paragraph or three, the first page—first. Let's see what happens if I write a few leads for that roommate column:

> As our grandchildren go off college for the first time, those of us who have the same roommate for forty, fifty years or more should share our co-habitation wisdom.

> ***

> Selective vision—or elective blindness—is the first quality a student should develop in facing a college roommate for the first time.
> I do not see the ironing we brought from New Jersey in 1963 that adorns one corner of our bedroom, and Minnie Mae, of course, will eventually learn not mention my . . .

> ***

> My first college roommate and I got along after we had a genuine, male, prancing-around, dirty-words fist fight in our closet of a room.
> Now, seeing freshmen arrive in cars hung with furniture, I realize their adjustment won't be to calculus, rhetoric, the philosophy of Hegal, but to fitting into a small room with a stranger who will get larger, louder, more difficult every day.

> ***

> I've heard of people who keep in touch with their college roommates decade by decade and I've heard of hostage victims who grow fond of their captors, but one of the good things about getting old is that I will never ever have to have a roommate again.
> I hope.

Perhaps in the nursing home, but I won't talk about that yet. And I am happy to have a mate, but it is fortunate we live in nine rooms, not one.

***

The first great lesson of college is that someone with a sense of humor—or sadistic need to cause trouble—has locked you and your roommate into a small space for a long year.

Now I have had a roommate for forty-one years and we are still working out all those trivial issues of territory that are so important to the human animal. As a full-grown, white-bearded sage, I have some wise counsel for first-year college students who face the first test of university life—I am supposed to live with. . . .

***

~~The first great lesson of college is that someone with a sense of humor—or sadistic need to cause trouble—has locked you and your roommate into a small space for a long year.~~

~~Now~~ I have had a roommate for forty-one years and ~~we are still working out all those trivial issues of territory that are so important to the human animal. As a full-grown, white-bearded sage,~~ I have ~~some wise counsel~~ advice for first year college students who face their first ~~test of university life—I am supposed to live with. . . .~~ roommate.

What have I been doing? Playing my way into an essay, trying on beginning points, voices, ideas the way you try on clothes before a party. Each lead gives me a direction in which I might go. The entire piece of writing grows out of the beginning that establishes:

- the question in the reader's mind to be answered in the draft
- the authority of the writer to answer it
- the direction of the draft
- the pace of the writing
- the form
- the voice

But What About All the Other Good Stuff?

There are two kinds of good stuff. One kind can be used to support and advance the focus of the story, to clarify and communicate your meaning. The other is material that will draw the reader's mind away from your message.

Supporting Material

Kurt Vonnegut, the novelist, once said, "Don't put anything in a story that does not reveal character or advance the action." That's a good rule for nonfiction, from corporate memo to literary essay.

Every piece of information, every literary device, every line, and every word must support, develop and communicate the meaning. Each comma, verb, statistic, reference, descriptive detail, transition, summary sentence should relate in a direct way to the central tension of what is being written.

The melody by itself is hardly enough. The meaning, focussed and sharpened, needs all the supporting material to reveal its full significance and to make the reader react emotionally and intellectually.

Distracting Material

The material you have collected through research, and the thinking you have done through writing that must be cut from the draft, however, is not wasted. It is all money in the bank. You may not spend it on this draft, but it is there, to be drawn on in the future.

And in a way it is still in the draft, even after it has been cut. The marble that has been cut away from the statue made an essential contribution to the statue. It is there in the revealing.

When starting a book, John Steinbeck used to write on one side of a single three-by-five card, the potential meaning of the draft. He would change it as he wrote and refined the meaning, but it gave him a sense of destination. That is a valuable revision device. Try it. Read your first draft and then write in *one sentence* the meaning you have discovered that you want to develop and communicate through revision. Put it at the top of the draft you are going to revise.

■ FRAME YOUR MEANING ■

You may not know much about writing, but I bet you know how to frame a picture with a camera. If you want to reveal the tranquil beauty of a flowery meadow, make sure you have not included the blur of traffic on the highway beside the meadow. If you want to comment on modern life by showing the cars rushing by unseeing, make sure you get both highway and meadow, and that you use an exposure that will show the cars blurring past the stationary beauty of the wildflowers in the meadow.

And that is what the writer does. For example, you might start an essay—or a short story—in either of these frames:

> It was not the commute between home and office and hospital that made it possible for him to survive but the ritual he followed every day when he pulled off the highway and studied the tranquil beauty of a meadow that had escaped the march of the malls. Some people liked water, but he drew strength from the ocean swell of land, the dance of wind on tall grass, the yearly explosion of wildflowers, each in its season.

***

> First the highway amputating this meadow from his grandfather's farm, then the subdivision behind the meadow, the strip mall on the left, the fast food place on the right, the hundreds upon hundreds of people who raced by this last meadow, not seeing the way it changed its color, hour-by-hour, under wind and sky, ignoring its tranquil beauty undisturbed until the annual explosion of wildflowers, promiscuous, profligate, so much more necessary to man than highways and burgers and cheap clothes and houses decorated with plastic possessions.

It helps us to realize how much we know about writing, without knowing we know it, when we use one art or craft and adapt its lessons, in this case focus, to our writing. Writing and re-writing go better when we can face our tasks with confidence. And we may not feel confident about our writing, but many of us feel confident with a camera in hand.

What to Leave Out

Draw a frame around the subject. You can even do this physically, by scanning the draft and drawing a line through everything that has to go. Sometimes I circle the material and mark it with an arrow heading off the page, or a question mark. Most times the decision is easy, but sometimes I have to scan the draft a number of times to see if it should go or remain.

What to Keep In

Keep in what moves your meaning forward. The remaining material must develop and communicate the focussed meaning.

"But almost everything went. My draft has shrunk!"

Good. This is an important stage in the writing process. When we have found our focus, it gives us space for complete development. Maybe half the material can be saved, or only a third, or a quarter—a page? Now you have room for the material you will add during the revision process.

Revision for the experienced writer is often a far more radical process than the inexperienced writer imagines. The experienced writer knows that to cut often means to reveal. Once I cut 237 pages from a draft in response to a modest but perceptive comment by a horrified editor. It *did* improve the book. The master writer knows that what is taken out is necessary to get the draft to the place where it can be made to work. The writing that is cut isn't a sign of failure but progress toward an effective final draft.

Most of us write a first draft that skims over the surface of our topic. That's appropriate. We are searching for our focus, our meaning. Once we find it, then a lot of the material we included can be jettisoned. Over the side with it. No regrets.

───────────────── ∎ *Writing Exercise* ∎ ─────────────────

Make a sketch—of course you can draw well enough for that—of what the reader should see when reading your draft. You may want to make a floor plan that will show where the camera is, what it can show and what it cannot. Can you do this with an idea? Absolutely.

Of course. Then it is a diagram that shows what is included—and excluded—in an essay on the French Revolution, in a plan for a new product due in a marketing course, in a report on environmental waste.

▪ SET THE DISTANCE ▪

An important issue relating to focus that is rarely discussed in most textbooks is the matter of distance. We can stand nose-to-nose with our subject or back off and see it from a mountain top, a space shuttle or even, by writing from a historical perspective, from a distance of hundreds of years.

The focus, as in a camera, depends on how far away you—*and the reader*—stand from the subject. The distance cannot be set by a rule book but depends on the subject, your purpose in writing about the subject, and the reader.

When to Use Close-Ups

Close-ups bring immediacy. We do not photograph the field of spring flowers but move in on one poppy, show the whole blossom or go even closer to a petal, perhaps catching a bee stopping by for breakfast.

In the close-up we don't look at the entire government but the legislature, not the House of Representatives and the Senate but the Senate only, not the committee structure but a committee, not the committee but a single senator, not the career of the senator but one revealing vote.

We can, with the close-up, expose the details of a scene, a law, a scientific experiment, a crime, a vote. We can reveal complexity and simplicity, cause and effect, action and reaction.

When to Step Back

We step back when we take a snapshot of a single wild flower then show the fields of wild flowers that stretch for miles to the Rockies. We show the historical and theological roots of the abortion issue and the pressures that caused the senator to take a position and now reverse it.

The distance shot allows us to make the generalization from the documentation; to put the anecdote, the quotation, the statistic, the scientific discovery in perspective. We can show what came before and predict what will follow. We provide a context. But what if I need to do both?

When to Zoom

You have a zoom lens. Use it. Don't just stand in one spot and try to see what's going on from there. Take your reader in close for emphasis, for clarity, for dramatic effect, to make the reader think and feel. Then zoom back so the reader understands the full implications of what the reader has been shown. In every war, photographers and writers use zoom techniques to reveal the landscape of war; then move in close to see the wounded or the dead, which reveals the price of war in individual human terms; then zoom out again to show the larger context of war.

Don't move in and back wildly, without purpose, like Uncle Max with his new video camera at the wedding. Move smoothly in close. Draw back halfway—in close once more—then back again in a pattern that serves the reader, giving the reader the information and the experience the reader needs to become involved in your subject.

———————————— ■ *Writing Exercise* ■ ————————————

Take that sketch, floor plan, or diagram you did before and move the camera further back or in close or concentrate the diagram or make it even more distant to see how that one element can change the impact of the text.

■ THE IMPORTANCE OF FOCUS ■

Drafting is putting in, but revision is the craft of selection. Revising a draft takes hundreds of executive selections. Each word may affect the meaning of the word next to it and words a page away; each line has implications for other lines; each sentence and paragraph changes the emphasis, pace, impact of other sentences and paragraphs; each piece of information influences the value of other pieces of information.

How do you make these decisions?

By using focus. Every decision is resolved by its value to developing and communicating your focussed meaning.

————————— ■ *Writing Exercise* ■ —————————

Select one specific in a piece of writing you plan to do. Focus on that specific and list what must be included in the final draft if you make that specific central. Change to another specific and see what in your previous list isn't necessary, and then list the new specifics that now become necessary.

STUDENT CASE HISTORY

Diane Soper

■══════■

First a note about the student case histories. They are real, undoctored case histories, written by freshman English students. They are not meant to demonstrate good or bad writing but to illuminate the actual writing process of students using this text. You will see many errors of grammar, usage, mechanics, spelling, punctuation, rhetoric, craft in these examples. It is one of the skills of revision to be able to read over the problems that must be dealt with in the last stages of the writing process, to identify those problems that must be solved earlier and in sequence.

None of the student papers came from a traditional, specific assignment. In all but the last case history, students were to write at least one personal essay, one argument with a strong point of view, and another paper that had a minimum of two or three outside sources. The topics grew out of in-class activities—including the students' listing of topics on which they were authorities, brainstorming and free writing—and individual weekly conferences with the instructor.

Diane Soper has problems—serious problems—with sentence structure, but before she can deal with those problems, she has to find her focus and, in this case history, you will be taken into the writing room as she discovers an effective way of focussing her paper.

Of her revision process, Diane Soper writes, "I let go in my first draft, with my thoughts and feelings. Then read it over and tried to focus on what I was really trying to say." She decided that she needed to focus on the subject. "In the first draft, I wanted to say something about myself. I realized—with revision—I wanted to talk more about the main character."

Soper added, "What pulled it all together for me was to look at the whole story like it was flexible, like it could move. The paragraphs that I took out were not thrown away. I took ideas from them, and got my point out in another way."

The day begins at 5:30 in the morning with a buzzing sound. You get up and hit the snooze button hoping to buy a few more precious minutes of sleep. The thought of getting ready for work slowly becomes an important goal. Time is not standing still, the alarm is buzzing again for the second time and you have to get up. "Get the kids up now!" They need to be at the sitter's by 6:00, if I am ever going to make it to work on time. The children are ready, and you stop to think if there is anything you might have forgot to do for yourself. If there is, it could wait till you got home. Have a nice day, be good in school and I love you, all sealed with a kiss. Now it is time for Mom to go to work. On the way to work you try to get that last sip of coffee in before it gets too cold. A few thoughts about how different your day might be from yesterday. Will today be as frustrating as yesterday? Or will it be more rewarding than you could ever think? Then again it might be just another day.

Remember we are publishing portions of these student papers as they were written. We are not doing this to expose the writer but because it is important for beginning writers to see real writing in its earliest stages. It is important for me, as a professional writer, to write badly if I am to write well. I must feel my way into what I have to say. There are many problems with the language in the previous paragraph, but it is doing its job; the paragraph is leading the writer toward a subject—in this case a subject the writer will choose not to pursue. Another time she may write a good paper about being a working mother. But as we will see, as she writes on, she will find the subject she wants to explore in this essay.

> The world of occupational therapy has many sides to it besides the politics of anatomy. There are inner thoughts and feeling that we sometimes take for granted. The clients feelings towards the person they work with every day, morning noon and night. The person who is there for them, to help them out of bed, into the wheelchair and then off to the bathroom, so they can learn how to brush there teeth again. The thoughts that must go through the clients mind as they become dependent on you to teach them independence. When will seven o,clock ever come he thinks to himself as he is lying in that ugly steel bed. Looking up at the ceiling waiting for you to come in that door. Outside of the door he can here voices, not wanting to cry out to them, because he knows they are all to busy and they'll say something like. "I will be wright back, in a minute." Then the minute turns to ten minutes, as if time had no meaning to him. He never needed to answer to time when he got up to get ready for school or work. But the car accident changed time for him. Now he waits for our familiar face to poke threw that door as it has every morning for the past six months. That funny little saying, you say to him in an Arnold Scwartzknicker voice; "We are here to pump you up." As I come in the room I still remember to keep that staff to client respect, by taping on the door as I walk into the room.
> The clients name is Joe he is twenty one years old now, he was eighteen when he got into a car accident going to school one morning with his brother . . .

Again there are many errors: misspellings, ungrammatical usages, incorrect punctuation, poor word choice, incomplete sentences, inappropriate order; and all of these problems will have to be dealt with in later drafts. But now we have a writer searching for her meaning, and she is moving toward it. We can see the writing taking her toward her subject—the world of rehabilitation experienced by an individual patient.

The second draft begins:

> As he is lying in that ugly steel bed. Looking up at the ceiling waiting for me to come to the door . . .

The writer now knows her subject. Her fourth draft follows. Each of us, of course, may see things we would correct or change if we were to make this piece our own. But it is not our piece of writing. Diane Soper may choose to go on and revise it more, or she may not. But she has learned how to find and develop a focus—a major breakthrough.

Here is her final draft:

> As he is lying in that ugly, steel bed, looking up at the ceiling, waiting for me to come to the door, outside the door he can hear voices. Not wanting to cry out to them because he knows they are all too busy and they'll just say something like, "I'll be right back in a minute." Then the minute turns to ten minutes. As if time had no meaning to him. Before he never needed to answer to time when he got up to get ready for school or work. The car accident that he was in changed time for him. Now he waits for my familiar face to poke through the door. As it has every morning for the past six months. That funny little saying I would say to him—in an Arnold Schwartzenneggar voice, "I am here to pump! you up!"
>
> As I walk into the room I still remember to keep that staff-to-client respect by tapping on the door as I enter. I walk into the darkened room to smell the stale nighttime air. Streams of light trying to peek through the drawn curtains. My first thought is only to get rid of the blanket of darkness that covers his room. I draw open the curtains with a, "Good morning, Joe."
>
> Joe is twenty-one years old. He was eighteen when he got into the car accident going to school one morning with his brother. The car he was driving was hit head on. Joe and his brother were rushed to the hospital. His brother had just a four-day stay, while Joe was checked in for a long, quiet three and a half months.

Joe was released to his parents' care. Comatose for three months and waking up with the lost movement of his left side. Still having the use of his weakened right. With a lot of work Joe's parents would teach the right side what the left once did. They both worked with Joe in their home day and night. I can't imagine the pain they must have felt caring for their son.

Joe is six feet two and weighs two hundred and two pounds. To look at him lying on his back you could only imagine what a great football player he was in high school. Now Joe is relying on a one hundred and twenty-five-pound life skills aide to get him up, dressed and off to the bathroom for his daily rituals.

Working as a life skills aide at a neuro-rehabilitation center to assist with the daily needs of patients; clients. Clients who have been injured in an automobile accident or who have sustained a serious injury to the head. I see each day how frustrating it can be for them. One needs to have patience and kindness. Then, most important something of a sense of humor.

In the morning when I enter Joe's room he pretends to be sleeping, not wanting me to know how much he anticipates my morning presence. "Good morning," I say again. Joe moves his right arm above his head and stretches out his right leg. With the miracle of reflexes the left arm follows but does not return back on the bed with his right side. His arm still hanging around the top of his pillow just above his head. I would move Joe's left arm back down to his side. "What are you doing, a cheer?" I would say to him. Joe would always laugh. He kind of had that Arnold Horshack laugh. He would get me laughing too!

How very often we take our own knowledge for granted. What if everything you knew were suddenly taken away? You're left only with a body that does not remember the simplest motions and those words. I can't do it!

I would work on his dissatisfaction. Show him the good behind these small movements.

It would take me almost an hour to help Joe get dressed and ready for breakfast. Joe would help me with the hard work, like moving him from the bed to the wheel chair. I would show him ways to get his shirt on over his head. Then off to the bathroom to wash up. At first he didn't want anything to do with brushing his teeth until I convinced him that if he didn't brush them he would look like Gumby. Joe did really well with brushing. Then with rinsing he would have a little problem of choking. I would have to coach him each time by reminding him which side of his mouth to hold the water on. The muscles on his left side wouldn't help him with a simple swallow. I would finish with a slick up combing of the hair for Joe and tell him how good he looked. Then it was off to breakfast.

When Joe eats he needs a specially adapted utensil to suit his needs. A fork or spoon with a larger grip attached. He needed a lot of his favorite foods cut up. Especially cheese steak subs, pizza and chicken. To sit back and eat freely while I work with Joe didn't come easy. I had to re-teach myself. I would have to watch and help only when it was needed, but very carefully, not wanting to offend him. Joe would say if I did offend him, "I'm not a baby." That would be my cue. I'm a mother of three children. It was not going to be like raising a child, where the mind is fresh and willing to accept guidance.

This was different—Joe remembers how to eat, to brush his teeth. He remembers all the things he could do. Now they are all nearly an impossible task for him. Joe would just as soon starve than to have me feed him pizza. I would give a little head start on eating by putting the pizza on the fork for him. The rest was up to him.

Four hours a day with Joe just on physical therapy. He would never be late for class. I would help him down to the physical therapy room. There would be no time to stop and talk. No! Joe was a worker. It almost felt like he needed every second of the day. As if it was not going to be there tomorrow for him. He needed his time and he was not going to share it. Other clients shared the P.T. room and the physical therapist. Joe wanted all of her attention. But with time he soon realized that we all need to share our time. He settled for my assistance.

He was in such a hurry for the tilt table: a table that was motorized. The table would tilt upward and had three velcro straps on it. Joe would have to be transferred from his wheel chair to the table and then strapped in. He hated this table at first. Maybe because he kind of resembled Frankenstein. It soon became another tool that he became dependant on. He knew it was helping him. He was really frightened of his appearance on the table.

I would see that Joe's ability to stand on his own with no assistance was very hard. I took for granted how we use our neck muscles, our arms, spine, hips, legs and feet.

After a while he did get used to it and knew that the table was helping his leg muscles, tissues and the flow of blood through his veins. Joe would start yelling at me when it was time for him to take a break. Joe knew he could only stay up on this table for a maximum ten minutes. Anything over this would be too much for the blood vessels in his leg. It was a battle to get him down.

Joe was given the control to the tilt table. A mirror was placed in front of him. Now Joe could watch his legs and if they started to turn a purplish color he would see to it that it was time for a break.

Life, living, do it all on your own. You're an adult now! As a child we are taught to do everything. We learn how to do the simplest task: combing our hair, putting on a shirt, drinking from a cup. Our teachers, parents, friends all guide us through this learning world. I was one of Joe's guides through this rough time and Joe was one of mine. I learned a lot about myself by working with him. He was a special learning experience in my life and inspired me to pursue a career in occupational therapy.

My editorial hand itches to attack Soper's essay and make it mine. She has a great deal to learn about the sentence but she writes so well without that knowledge that I want to grab hold of her almost-sentences and train them to *my* voice and to *my* meaning. And that is the bigger temptation. I grew up with a paralyzed grandmother and have published a novel about a quadriplegic. I want to make her story mine, but that would not help her and it would be a felony. She has to discover her story by learning to use her language with greater skill and by continuing to look at her subject with her own tough, compassionate, honest eyes.

Writing gives you the opportunity to find the focus of those stories that illuminate the many worlds in which we live. Then you can follow the sentence to the stories' meaning—the meaning of the life you have led and are leading.

CHAPTER 5
RE-WRITE TO COLLECT

Writers don't write with words.

Writers write with information—accurate, specific, significant information. It is essential to the craft of revision to consider the information communicated in the draft. Words are the symbols for information, and when there is no information behind the words the draft is like a check with no money in the account: worthless.

Effective writing serves the reader information the reader can use to understand the world, to think more clearly, to make better decisions, to learn, to act, to appreciate and enjoy life, to become an authority in the eyes of those around the reader. The list of reasons readers want and need information is long. Effective writing is constructed from sturdy bits of information.

Beginning writers can easily understand the necessity for specific information in non-fiction articles designed to communicate information—the research paper, laboratory report, minutes of a meeting—but less understanding of the need for information when writing about ideas or feelings. Yet specific information is absolutely essential in both forms of writing.

Which of the following examples from a history of World War II would you read?

Tech-strategists retroactively engaged in cognitive studies of an interdisciplinary nature theorized that the late entry of United States forces, the very lack of preparedness that was so castigated was, in contradiction to all expectation, a situation that produced positive, if unexpected, tactical and strategic results.

OR

Hitler was prepared for war, and when his Stukas and Messer-schmidsts overran the low countries of Northern Europe, the U.S. Air Corps could not have flown with them even if the U.S. was in the war. Ironically, those who mobilize last will enter combat with the "latest" equipment. If the enemy can be held at bay, the later entry will fly planes, as we did, that are technologically better than the German planes that seemed invincible a year or two earlier. Some scholars have argued that poor preparation is the best preparation for war.

I try to write of my feelings about being unable to help my 20-year-old daughter when she became ill and died. Which example is more effective?

A parent always wants to protect a child and never, no matter how irrational it is, stops feeling guilty if a child is killed or dies from an illness, feeling there must have been something the parent could have done.

OR

Lee

Remember me not
when I was kept from you
in the waiting room, not
when I sat in an office signing
your dying, not
when I pushed you on the swing
higher than you had ever flown
and you looked back as I grew small,
certain I would always be able
to save you.

Effective writers of non-fiction, fiction and poetry do not tell the reader how to feel but give the reader the specific information to make the reader feel.

■ DIAGNOSIS: TOO LITTLE INFORMATION ■

Quickly read through your draft and put a check in the left-hand margin whenever you give the reader specific, accurate information.

Some of the forms of information you may see, depending on your subject, form, and audience are:

revealing detail, fact, statistic, direct quotation, anecdote, first-hand observation, precise definition, attribution, and authoritative citation or reference.

Information may take many forms, of course, and may be incorporated into the text. A book of writing might be written on the basis of composition research reports and have many foot-notes and other forms of attribution; this textbook is written from personal experience and has few references to other sources. Neither way is right or wrong; each is appropriate to the forms of writing.

There is no quota on the amount of information needed in a given piece of writing, but if you have pages or paragraphs without checks, then you should read the draft to see if *the reader* needs more specific information.

────────────── ■ *Writing Exercise* ■ ──────────────

Read through your draft slowly, marking the draft wherever the reader would appreciate a piece of specific information. Then collect the information the reader needs and insert it in the draft.

■ THE IMPORTANCE OF INFORMATION ■

It is normal for inexperienced writers—and some not so inexperienced—to become infatuated with words. The condition is called "word drunk." The writer staggers down the page, spouting words that may, accidentally, sound wonderful but say nothing. Most readers do not want a word-drunk writer any more than they want a shaky-handed surgeon.

Provides Reader Satisfaction

Readers are hungry for information. They want the images and facts, revealing details and interesting quotations, amazing statistics and insights that make them see, feel and know their world better than they did before the reading.

It helps to understand the reader's desire for information when you realize that one reason you read is to become an authority.

The information-rich writer makes the reader an authority, and the reader in turn, becomes an authority. The reader broadcasts new information to family, colleagues, friends, and gains status in the process. Joe is in the know; Belinda is in the loop. A good piece of writing ignites a chain reaction of communication.

Most of the writing that satisfies us as readers has served us an abundance of information.

Establishes Authority

Readers believe specific information. If you want to lie, lie with statistics. Precise information makes the reader believe that you know your stuff. But serve up one piece of precise information that the reader knows is wrong and the reader won't believe anything in your draft.

Accurate information is the reason that readers trust the writer; inaccurate information is the reason that readers mistrust the writer. Readers test the draft by noting the information that relates to their world. When I read something about a newspaper or a university, I am especially critical. If I think the writer is on target, I trust the writer's comments about institutions I do not know well; but if the writer's comments do not fit my knowledge of newspapers or universities, then I suspect everything the writer says.

Produces Lively Writing

"How can I make my writing lively? It is dull, dull, dull."

Yesterday I was asked that question and I answered, "By concrete details more than anything else. Lively writing is specific, not vague, abstract and general. It builds the generalizations on the page and in the reader's mind from specific pieces of information that surprise and delight the reader."

─────────── ■ *Writing Exercise* ■ ───────────

Take a newspaper or magazine story about a subject you know—surfing, working in a supermarket, life in a sorority, living in Chicago—and read it, circling the specific pieces of information you know are true. Cross out the ones you know are untrue or questionable. Then read the piece again to see the effect of your trust in or distrust of the writer.

■ THE QUALITIES OF EFFECTIVE INFORMATION ■

Effective information is information the reader uses successfully. A manual tells the reader how to solve a problem with computer software; an editorial makes the reader change a vote; a biography puts an historical figure in perspective; a literature text illuminates a poem; the poem makes the reader see the woods the reader passes every day, with increased perception.

Accuracy

One of the reasons readers respect specific information is that the author takes the risk of being specific. Much of what pours through our ears is purposely vague, general, abstract; the writer uses politic language that can't be nailed down, information for which the writer is not accountable. Specific writing is unusual, and readers like it, *but* it must be correct. One slip, as we have said, and the reader will not trust anything the writer says.

Factual Accuracy

The first accuracy is the truth of the fact: the number of miles in the marathon, the cost of the school bond issue, the day of the Battle of Gettysburg, the ingredients in the formula and their precise amounts, what the president actually said in the speech. All must be right.

These facts can be checked with authoritative sources and should be. Often the writer will check the source with another source; just because a fact is in print doesn't mean it is a fact.

Some questions that you can ask of a fact, then check and re-check if the answer is not "yes," are:

- Does it make any sense?
- Does it seem possible from what you know of the subject?
- Is it consistent with other facts you know are correct; does it fit the pattern?
- If this fact is true, does it change other facts in the draft—or the meaning of the draft?

Of course, the most interesting and significant facts will not get a simple "yes," but those pieces of information will have to face the skepticism of the reader. They deserve to be checked and re-checked.

Contextual Accuracy

The writer has the responsibility to make sure the information in the text is **accurate in context**. That is a far more difficult, and more important, task than just checking the number of pork pies at a church supper. Some questions that may help you test the contextual accuracy of a specific piece of information are:

- What does this detail mean to the reader? What message does it deliver to the reader?
- What does this specific make the reader think? Feel?
- What impression is conveyed by this specific and the information that surrounds it?

- What is the pattern of meaning being built by this piece of information and the specifics that come before and after it?
- Is this piece of information *accurate in context?*

You are using specific information to construct a meaning and you have the obligation to make that meaning true.

Specificity

We like specific information because it catches our eye, or ear; sets off chain reactions of memory or imagination; gives us something to play with; makes us think and feel.

The Disadvantages of General Information

Writers—and politicians, corporate executives, educational administrators, bureaucrats—often use generalizations to avoid responsibility. You can't nail a generalization down. Watch a presidential press conference. Either party will do.

Meanings, feelings, ideas, generalizations, theories, abstractions are all important. We use them in writing but we are fortunate when we can cause them to happen in the reader's mind because of what we have said on the page. The details are arranged in a pattern, and that pattern makes the reader construct a meaning or experience a feeling. We cannot usually construct a meaning from generalities. There is nothing there to cause a thought. There is just someone else's thought to be accepted or rejected. It is given to us without documentation. We can't see backstage to discover—and evaluate—how it was put together.

Undocumented generalizations—those that are not built out of accurate, checkable, specific information in front of your eyes—are weak, flabby, vague, dull. And the writing is . . .

The Advantages of Revealing Details

A revealing detail is a specific that says more than one would expect. "The additive may cause genetic problems" sounds specific, but notice the difference when you read "Urgatative, a normal food additive, may cause nerve damage in a pregnant woman and her child—and her child's child." Revealing details expose the subject; they connect with other details to construct an opinion, argument, theory, poem, story, report that can be studied, challenged, tested.

The details themselves have power. *"The mayor won"* is not equal to *"The mayor won five votes to her opponent's one."* And details can make the reader think beyond the end of the sentence: *"The first African American to be elected mayor won with seventy-one percent of the white vote."* Or make the reader feel: *"In her victory statement she thanked her husband, her children, her parents, her campaign workers and then lifted Soozie, her seeing-eye dog, up on her hind legs so she could acknowledge the cheers of the mayor's supporters."*

Significance

Effective information is significant. If the new mayor had a dog named Soozie that would mean nothing, but the fact is that a seeing-eye dog makes a statement about the possibilities for the handicapped—a term the mayor never uses—and implies that her administration will pay attention to that minority as well.

Resonance

A powerful detail has resonance; it gives off an explosion of implication in the reader's mind. Resonance causes the reader to begin thinking, taking the text of the page and exploring its implications. The mayor reminds the audience that she went to school in this city, but that although each year the white high school students elected a mayor who served for a day in City Hall, she was *not* allowed to serve when her African American classmates elected her their mayor.

That piece of information resonates. The reader can imagine how she felt then and how she feels now; can imagine the struggle that put her where she is now; can imagine how many other talented citizens did not and do not have the chance to serve; can imagine the changes that allowed this to happen and the changes that need to take place.

Connection

The information you choose should make significant connections in the writer's mind and in the reader's.

With Topic. The information should relate to the topic and advance its meaning in the reader's mind. Often writers collect interesting information and they can't let go of it. The more interesting it is, the more it draws the reader's mind away from the message you are delivering. Each specific should amplify, clarify, extend the topic.

With Other Information. The specific details should work with the other pieces of information in the draft. They should build toward an increasing understanding of the meaning that your draft is delivering to the reader. Some details will increase the impact of the other details; other specifics will qualify or limit the meaning that is building up. During revision you have the opportunity to check on these relationships.

With Reader. The information you choose to use should relate to the reader. Both the type of information—statistics, anecdotes, quotations—and the meaning the information bears should appeal to the reader. What will persuade one reader may turn another off. It is your job to select appropriate information from which to construct your draft.

Fairness

The best way to evaluate the fairness of a draft is to stand back and become a person in the piece of writing or someone who is affected by it. What would you think if you were in the piece or if your reputation were affected by its publication?

───────────────── ■ *Writing Exercise* ■ ─────────────────

To reveal to yourself the power of specific information, write a general statement about someone you know—"*Jennifer is a nice person*"—and then list twenty specifics that document her niceness. Now change the statement by adding "*not*" before "*a nice person.*" Write down another twenty specifics. Now write a paragraph that is fair, documenting the strengths and weaknesses of the person with specific, accurate details.

■ BASIC FORMS OF INFORMATION ■

We deal with information all the time, but we don't think of it as information we might use to construct a piece of writing. It may be a good idea to remind ourselves of the common forms of information.

Fact: A precise piece of information that can be documented by independent sources. Linda J. Stone was *elected governor.*

Statistic: A numerical fact. The Governor received *5,476,221* votes.

Quotation: The direct statement by an individual speaker or writer or from a document, book, article, report. "*The first Monday of every month the door to my office will be open to any citizen who wishes to see me.*"

Anecdote: A brief story or narrative that makes a point that documents the point a writer is making. The parables, such as the story of the loaves and fishes in the Bible, are a form of the anecdote. "*Governor-elect Stone drove the family car, a five-year-old Ford station wagon with 137,422 miles on it, to election headquarters herself.*"

Descriptive Detail: A specific that reveals a person, place or event. She held high the "*old fashioned wooden clip board*" she carried during the campaign on which she had made notes on what individual citizens told her during the morning coffee meetings she held in private homes, offices, schools and factories.

Authoritative Report: A document that is accepted as accurate by a responsible organization such as a court, corporation or academic discipline. In her acceptance speech, Governor-elect Stone cited the *"Newkirk Report that called for small classes in the public school system and a paid-by-the-state system of summer retraining for teachers."*

Common Information: Information the reader can be assumed to know—three strikes and you're out; the president outranks the vice-president. *"The governor-elect of this state is immediately given a state police driver and an official limousine."*

These are just a few of the most common forms of information. There are obviously many more, and I have used a journalistic example common to all of us. Each academic discipline, each corporation, each profession, each government agency will have its own basic inventory of information forms. Most of them will, however, fit these patterns. The literary scholar will have many quotations from the work studied and from its critics; the economists will cite many statistics; the physicist will have many facts, mostly in the language of mathematics; the environmentalist will build a draft with many descriptive details; the historian will make good use of authoritative reports; the social case worker may have a great deal of anecdotal evidence. In every case, however, the most effective pieces of writing will be constructed with specific information.

■ WHERE DO YOU FIND INFORMATION? ■

One of the reasons I am glad to be a writer is that I am forced to continue to learn. I have to mine my world for the specifics I can use to discover what I have to say. The process of writing is a process of thinking, but if the thinking is to be effective—and if readers are going to read and use it—it must be built from information. And to get that information I have to study my subject.

Memory

We fear that writing will prove us ignorant. But writers discover how much we know by writing. I don't think I remember what it was like when my father, during the Depression, was fired again and again. Then I write about one of the times and begin to hear him blaming others, never himself, bragging how he told his boss off, how the reason he was fired was that he would be replaced by a "college man." I find out that I remember more than I thought I would and I also realize I can now begin to re-create my doubts and questions about my father, how I felt, how I would try to live my life. Writing reveals how much I know. Of course, writing may reveal my childhood but it will never reveal nuclear physics, no matter how many drafts I create. Drafts reveal the knowledge of which you were unaware, not the knowledge you do not possess.

If your draft did not show you how much you knew, you may need to take a step back and discover what you know that you didn't know you knew.

Before Writing

I start withdrawing information from my memory bank by brainstorming a list, putting down everything I know about the topic I am going to research. Or I fastwrite a "what-I-know" draft to surprise myself.

During Writing

While I am writing the draft, I encourage the connections I do not expect by writing fast—without criticism. I continuously make use of my writer's memory, which is stimulated by the writing act, making me aware of what I know that I didn't remember that I knew. I try not to know too well what I may write, allowing the discovery drafts to lead me. I also may make notes in the draft or on a pad of paper beside my computer of things I suddenly discover in the writing: references to other writing, connections between facts or ideas, new patterns of meaning, sources I have to explore.

Observation

In the academic world, observation is often overlooked, yet it can be a productive source of significant information. If you are writing a paper on criminology, visit a police station or a jail; on health care, spend a few hours sitting in a hospital waiting room; on economics, walk through a supermarket or a mall; on literary studies, browse through that part of the library in which your subject is preserved; on government, attend a meeting of the school committee, town or city council.

When you observe, make notes. That activity will make you see more carefully as well as preserve what you observe. Note your first impression of a place, a book, a person. Make notes on what is and what is not; what is as you expected and what is not. Look for revealing details: how people interact, where things are placed, what is happening. Use all your senses: sight, hearing, smell, taste, touch. Take account of how you feel, react.

Interview

Live sources should not be overlooked. If you are writing about schools, interview students present and past as well as teachers, administrators, schoolboard members, parents. Read the books and articles about schools, but also go to see the people in the classroom.

Good interviewers are good listeners. Few of us can turn away from a quiet, receptive listener who makes us an authority by asking our opinions. Try not to ask questions that can be answered with a simple "yes" or "no." Not "Will you vote to make professors sing all their lectures?" but "Why do you think it is important that professors sing their lectures?"

I like to prepare for an interview by listing the questions the reader will ask—and expect to have answered. There are usually five questions—give or take one—that must be answered if the reader is to be satisfied.

"Why is tuition being doubled?"
"How will the money be spent?"

"How do you expect it to affect students?"

"What are you doing to help students who cannot pay?"

Listen to the questions and follow up on the answers to your questions: "We are raising tuition because the faculty is underpaid." "What evidence do you have that the faculty is underpaid? Can you name faculty members who have left because of pay? What positions are unfilled because the pay is so low?"

Note the answers to your questions, but check the ones you have any hesitation about or the ones that are most dramatic and surprising: "We are raising tuition so we can build the first class ping-pong stadium our students demand." Journalists rarely check back with the person they interviewed, but I usually did. The purpose of the interview is not to trick the person being interviewed but to get accurate information to deliver to the reader.

Library

We live in an increasingly complex world. We need information on toxic wastes, traditions in Italian politics, arctic survival techniques tested in Siberia and Canada, Norwegian exchange rates, the development of a new strain of AIDS virus in Africa and a treatment development in Paris, the translation of the work of a Nobel prize winner who writes in Arabic. The list goes on and on. Libraries are the intellectual closets of humankind where information is stored until we need it.

Search Techniques

Can we find the information? Not in my closet, I find myself answering. Fortunately, humankind has librarians who organize information so that it can be recovered.

Your library will be changing and increasing its access to sources of information. Go to the reference desk, but do not only ask them to help you get information on your topic; ask for instruction in how to use the library so that you can do your own research. That is a skill you'll need as a lawyer, salesperson, police officer, politician, social worker, doctor, scientist, teacher, nurse, retail store manager.

The starting point for all the resources available to you—books, reference guides, monographs, articles, reports, audio and video tapes—is your librarian. Use the librarian to learn how you can tap into the abundance of information you need to draw on to write— and think—effectively.

Creating the Bibliography

The inexperienced library researcher pounces on an enticing piece of information and forgets to note where it came from. It is useless. It is no longer a piece of important information if it cannot be cited with an accurate attribution and checked by you and a reader.

Use a system of file cards—usually organized by topic—and within that, alphabetically by author, to record the complete title of your source, the author, the publisher, the copyright or publication date, the edition, the library reference number, the page on which you found the reference. Take down all the information that you or a reader may need and record that information in the way demanded by your discipline or in a way that will best serve anyone trying to find and use that reference.

It is hard for those who have not used scholarly materials to understand the necessity of footnotes and bibliographies. It is not just to establish the authority of the writer, but to serve the reader who is doing research, to let that interested reader follow the trail of scholarship that led to your writing. It is a matter of more than etiquette. Footnote references and bibliography are a duty if you are to participate in the intellectual world, adding your knowledge to those that went before, so that it is possible for those who come after to build on your contributions.

Plagiarism. Plagiarism is using someone else's ideas or writing as if they were your own. Plagiarism is theft.

But plagiarism can be avoided if the writer takes accurate notes that indicate precisely what words are the author being cited and which are the words of the researcher summarizing what the author said.

Of course, many students do not know the difference between quote and paraphrase. Here is the difference:

Quote is using someone else's exact words. Those words are enclosed in quotation marks: "*President Franklin Delano Roosevelt said, 'The only thing we have to fear is fear itself' in his 1933 inaugural address.*" They are attributed to the person who said them by a direct attribution in the text, a footnote, or both.

Paraphrase is putting someone else's idea in your own words. "*President Franklin Delano Roosevelt spoke of the insidious effect of fear on the nation in his 1933 inaugural address.*"

Attribution

Attribution connects information with its source.

No attribution: "*Alcoholism is a problem on the campus.*"

Attribution: "*Alcoholism is a problem on the campus according to Chief Ruth Grimes of the University Police, Chief Marvin Wiss of Campustown Police, Doctor Irving Feld of the University Health Center, Psychologist Mary Pratt of University Counseling Service.*"

The reader deserves to know where the information comes from: "Who sez?" The reader ought to be skeptical, questioning the text, challenging its authority. And we have the obligation to answer that challenge.

Unstated

All the information in the draft has an attribution. That information that has no footnote or reference is attributed to the writer. We must remember this and make the information we use on our own authority accurate, specific, and fair.

Stated

In formal, academic, research or scholarly writing, the attribution is provided by clear statements in the text, together with a footnote, or with a footnote alone. The footnote style may vary in a history, psychology, physics, zoology, mechanical engineering course. Make sure you know the style that each instructor expects.

Should your chosen style be less formal, it is still important to provide, in the text itself, a reference—"*as President George Bush said in his televised report to the nation on the action in Panama,*"

December 20, 1989. . . ."—that will allow the reader to look up the original document.

Effective Note Taking

It is important to have a system of note taking so that you have accurate, readable (you ought to see *my* handwriting) notes with the source clearly indicated.

File Cards. Most writers find that the most efficient system involves note cards—three inches by five, four-by-six, or five-by-eight—on which most information can be placed, one quotation or piece of information to a card, together with the source. The cards can be easily carried to the library and ordered and re-ordered into categories as the research develops. It is easy to check back with cards and to line them up for reference while re-writing.

Computer Notes. More and more people are typing their notes into a computer using software programs that do a spectacular job of organizing, ordering and re-ordering, information.

Print-out, Photocopy, Fax. The electronic world in which we live has changed the way in which we collect and take notes. I do a great deal of photocopying when I research, making sure, however, that I write on each photocopied page where it came from. I either place these notes in a file folder—with the same heading as my computer file—or type them into the computer. I also get print-outs from computers or have a direct computer transfer of information into my computer files. Those are either placed in the stationary file or the computer file. Other material comes in by fax and I treat it the same way. It is vital that you know where each piece of information comes from at every stage of the research and writing process.

Reader-Granted Attribution

There is a third form of attribution, the one granted by the reader. When you write to a particular audience—police chiefs,

supermarket managers, colonial historians, college undergradu-
ates, town managers—you can refer to common experiences, express common opinions or frustrations, articulate the thoughts
and feelings of your audience, and if they recognize and agree
with what you are saying, that act grants you authority.

You may not have to attribute common knowledge, but you
should indicate the source of any surprising or unusual information.
This is a gray area, and I would lean over backwards, when in
doubt, to make sure your reader knows the source(s) of information
central to a piece of writing that depends on research.

─────────── ■ *Writing Exercise* ■ ───────────

List the sources you can contact (within the limits of time and geography) that might have the information you need to support the
meaning of your draft with accurate, specific information. Contact
them and acquire the information you need.

■ WRITING INFORMATION ■

Once the writer's information inventory is full, the challenge is to
use selected information gracefully and effectively.

The Craft of Selection

Writing—and every other art—involves the craft of selection. A
great deal of good material, information the writer worked hard to
collect, will be left out. That is the mark of a good piece of writing. And, in a sense, it is all there. The reader feels the abundance of
information behind the page. The draft is not thin. It has depth and
weight; it is worthy of attention.

Style

The style you will use to write will depend on the message or
information you have to convey, the reader to whom you are

writing, the occasion of writing, the publication in which the writing is to appear, the genre in which your message is carried, the voice of the draft; but there are some basic techniques to consider.

Word

The individual word carries information to the reader; and to write lively, information-laden prose we should make sure that each word carries an adequate load of information to the reader. The words that can carry the most information are nouns and verbs.

Each case depends on the situation, but consider the information contained in the simple words "house" and "home"—there's a choice worthy of consideration. Carry it further:

She lived in a house.	home.
	shack.
	mansion.
	hovel.
	palace.
	tenement.
	prefab.
	trailer.

And look at a simple, active verb:

He	walks	**into the house.**
	slams	
	strolls	
	darts	
	charges	
	saunters	
	dances	
	tiptoes	
	marches	
	clumps	

And the list can go on. Notice how much information a noun and a verb can convey in a simple sentence:

They	walked	into the	house.
	slammed		home.
	strolled		shack.
	darted		mansion.
	charged		hovel.
	sauntered		palace.
	danced		tenement.
	marched		prefab.
	clumped		trailer.

Now connect different nouns and verbs. Notice how you change the information and the message.

Phrase

The phrase, that fragment of language that is less than a sentence, is often an effective way to communicate information.

He walked the streets   **like a soldier on patrol.**

Now I'm in the game of writing: can I say the same thing by simply changing "walked" to "patrolled"?

in fear of each shadow

as if he owned them

listening to his own footsteps

seeking the safety of shadows

unaware of who was following him

Sentence

The simple sentence, as we have seen, can carry more information than we might expect, but we can load up the sentence with a great deal of information:

A university has been described, by Grayson Kirk, as a collection of colleges with a common parking problem, but I am struck, after returning to the university I attended, at how resilient an institution it is, welcoming unexpected thousands of veterans on the GI Bill after World War II and changing its curriculum, year after year, in response to society's need for study and research in many fields unheard of when I was an undergraduate: computer studies, space sciences, women's studies, black studies, environmental sciences.

Paragraph

I think of paragraphs as the trailer trucks of prose that carry a heavy load of information to the reader:

> I wonder if my post-World War II generation will be known as the generation of the single-family house. I dreamt of living in a single-family home as I was brought up in rented flats where I was shushed by the fear that the neighbors upstairs or down might hear. Hear what? Anything. Music, fights, kitchen clatter, the Red Sox game, bed springs, bathroom flushes, burps, curses, footsteps, doors shutting, drawers opening. And then an uncle bought a single-family home in east Milton and I knew that was what I wanted. A house of my own, far enough from the neighbors to thump a ball, yell back, play my music at proper volume, flush the john after midnight. And I made it, one of the veterans who gloried in urban, suburban, and exurban sprawl. But my daughters, both successful, making more money than I did at their age, may never be able to afford a single-family home.

There are many other ways to carry information to the reader including illustration and graphics, but most information is communicated by word, phrase, sentence and paragraph.

Remember that the reader is hungry for information. That is the principal reason we have readers. They want to read the specific, accurate, interesting information that will turn them into authorities on our subjects.

▪ REVISING ACADEMIC WRITING ▪

Academic writing has its own particular demands for information. The form of information needed depends on the discipline—even the discipline within the discipline. A paper in clinical psychology may need case histories, while a paper in experimental psychology may need graphs and statistics. Effective writing in every genre—from poetry to a laboratory science report—is constructed from the information that the reader needs.

The purpose of academic writing is not just to test the student's knowledge, although that is a consideration in school. Its main purpose is to inform and to persuade, to advance our understanding of black holes in physics, to assess the importance of the new records of Lincoln's law practice in history, to have us re-consider Faulkner's novels from a feminist perspective. Our knowledge moves forward on a trail of documented evidence that can be checked by those who follow the trail of old scholarship, which is extended by new researchers and scholars.

STUDENT CASE HISTORY

Glenn Bailey

■══════■

Glenn Bailey describes, with unusual candor, how he chose his topic, "I chose the subject of vegetarianism because I couldn't think of anything else. Actually a friend suggested it. It's not something I like to talk about, never mind write, but I did it anyway."

He identified a problem immediately, "by trying to figure out how I would put it across." He didn't do any research, just brainstormed, and then, "just splatted it down on the page." This is his discovery draft:

> I'm going to tell you why I think people should not eat meat. I'm not going to have an easy time of it, but here it goes. First and foremost, the animal has the right to live. Secondly, we don't have the right to decide if it should live or die. To me it is clear, to others I can only speculate. An unbelievable amount of animals die each year for human consumption and as far as I can see, it's completely unnecessary. It's not only unnecessary but absurd. Whether people are ignorant or just don't care I don't know. I'd like to think it's ignorance so I'll try to enlighten or at least explain my point of view. Like almost everyone I know, I was raised eating meat and never once thought it peculiar. As I got older I was exposed to some animal activist literature. I thought that the people involved were pathetic but

it made me think. Why do these animals die? Natural instinct told me that we need them for food, meat being one of the four food groups and all. Well as it turns out, we don't need them for food. I thought it would be strange if a person stopped eating meat. All these things that came into my head couldn't even come close to justifying the death of one of Gods' creatures. Over the next day or so, this subject would pop into my head every time something related would happen. Watching television and seeing people eat. After investigating more on the subject through magazines and such, I began to think that maybe what we're doing is wrong. Now I do believe it's wrong.

I recently read a paper that put my feelings into words and made this paper difficult to write. If you ever get the chance and want to understand what I'm saying better, I recommend you read a piece called "The Vegetarian Savage: Rousseau's Critique of Meat Eating," by David Boonin-Vail. The piece is an analysis of another piece written by Rousseau called "The Vegetarian Savage." After reading this paper, I felt as though anything that I wrote would only be an attempt to say just what he said.

In truth, we do not have to kill these animals for food and if people would think about it, I think things could change. I've been approached with a number of arguments against my view, but not one has ever made me believe it's our right to kill these creatures. I've been told that it's necessary for population control, but we breed animals strictly to eat. The ones we don't breed, well, who are we to decide whether they live or die? People have told me that they just could not give it up. That is certainly no reason for death. I think a lot of people are ignorant and lazy and afraid of change. I don't want this to be a paper that condemns everyone who doesn't see things my way, but it's obvious to me and I know no other way to explain myself without being so direct.

Bailey has done what many beginner writers have done. They simply state their thoughts and feelings. The reader knows how the writer feels, but is not persuaded. The reader needs documented evidence that may convince the reader to think differently. The writer was finally persuaded himself, through conference and workshop, to gather objective evidence and to attempt to build a persuasive argument with this kind of documentation.

This essay has a long way to go before it will persuade a reader opposed to being a vegetarian, but you can see the difference a few authorities have made. I think he needs more and better qualified authorities—and I am suspicious as to whether his quotations are particularly well chosen or even accurate. Several of them have grammatical errors that should not have appeared in print and at least one must have a word missing. Still, we see the important turn the writer has made by collecting specific documentation.

> The routine of animals being killed for food is an overlooked custom that people should examine more closely. We should look carefully at why we do it and learn the truth about popular myths. Once people forsake their blind faith of past tradition, maybe animals can receive some relief from this unjust treatment. I think we should share the right of life that humans have with cows and chickens and pigs alike.
>
> To find out some of the reasons why people don't see what seems so clear to me, I talked to my friends, family and people I work with. I asked them what they thought about all the killing. I was given three general reasons to justify it; it's man's nature, health and population (control).
>
> Education seemed to be the one thing most people lacked. Therefore, I'm committed to enlighten. Hopefully, through education, we can take a step towards stopping the slaughter.
>
> It's popularly believed that man is just 'meant to eat meat.' This 'we always have, we always will,' attitude is something I don't accept. Stephen R. L. Clark, a Professor of Philosophy writes, ". . . we *can* live well without meat and that *we* are not entirely custom-bound, even if the non-human carnivores are. The point at which our ancestors turned from the largely frugivorous diet of those ancestors we share with the apes is in dispute. . . . It does not follow, finally, that it is somehow obligatory, because 'natural', to eat flesh. . . . Our ancestors made a choice, for what·ever reason or upon whatever impulse of desire: we are not bound to imitate their errors. *Nature is not static."* (Clark 176–177) Because humans, for the most part, now, are meat-eating animals, doesn't lock us into this habit. Change is not an unusual thing, moreover quite natural given our past. A vegetarian diet is as natural, if not more classical, than one including meat.

Studies now show because of the physical make up of the human body that we may be by nature a vegetarian animal. One example is our teeth and the fact they are blunt, not sharp. Sharp teeth, like the cats, are a characteristic of a carnivorous animal and blunt of a vegetarian animal. Similar studies of the human organs show us to be non-carnivorous.

Health is another reason people disregard the idea of vegetarianism. It's believed that meat is an important part of a healthy diet. This is a popular misconception. It's becoming more widely known that a vegetarian diet is as healthy, if not more healthy, than a diet including meat. Peter Singer is a Professor of Philosophy and Director of the Centre for Human Bioethics at Monash University, Melbourne, Australia. He writes "Nutritional experts no longer dispute about whether animals flesh is essential; they now agree that it is not. If ordinary people still have misgivings about doing it, these misgivings are based on ignorance" (Singer 185). As for proof to back up this statement, he writes "Strict Hindus have been vegetarians for more than 2000 years. Ghandhi, a lifelong vegetarian, was nearly eighty when an assassins bullet ended his active life. In Britain, where there has now been an official vegetarian movement for more than 125 years, there are third and fourth generation vegetarians" (Singer 184–185). I feel that there is more than enough evidence to put to rest all doubts about whether or not vegetarianism is a health choice.

Population seems to be another popular argument. It's believed that population control of wild animals is important or they would go out of control. This justifying the hunting and eating of deer and alike. I recognize the problem, but I don't think it's up to the human race to decide how many of each species should be on the planet at a given time. Clark puts things in perspective when he writes, "There is, after all, a human population problem as well: should we argue that we must breed as many as possible (from selected lines) and control the numbers by imprisonment and war?" (Clark 61)

There are other alternatives to solving the problem of lack of space in the wilderness. Harriet Schleifer, co-founder of Quebec's Animal Liberation Collective, writes, "Wildlife conservation is a popular concern for many people, though few know the extent to which domestic animals compete with wildlife for space and resources . . . . more than half of the country's total land area, is presently used for meat, dairy and eggs operations, making it un-available as human or wildlife habitat" (In Defense . . . 67–68).

If people chose not to eat meat, it could have positive impacts on the work which most people don't realize. On world hunger, Singer explains that in the year 1968, 20 million tons of protein, most of which could have been consumed by humans, was fed to livestock (not including dairy cows) in the United States, producing only 2 million tons of eatable protein (Singer 171). The amount of food used to raise animals, as opposed to feeding people, is so sig-nificant that, by itself, it's reason enough to stop raising them for slaughter.

I realized that a major difference between my view and the views of many others is a moral one. I get the impression that peo-ple feel humans are superior to animals. People think humans are more intelligent than animals, therefore should be in control. When I think of this theory, I think of today society and see legalized slav-ery, only the most powerful people allowed any freedom.

I strongly believe that animals have as much right to this world as we do. David Boonin-Vail writes "The civilized man, that is, calls upon his reason to assure him that, since animals do not share in all of the human attributes, their pain must somehow be less painful, their distress somehow less distressing, their suffering somehow less real" (Boonin-Vail 78).

People must realize that animals have rights. The right to live and the right to roam, just as people do.

Works Cited

Boonin-Vail, David. "The Vegetarian Savage: Rousseau's Critique of Meat Eating." Rev. of multiple works by Jean-Jacques Rousseau. Environmental Ethics Spring 1993: 75–84.

Clark, Stephen R. L. The Moral Status of Animals. Oxford: Oxford University Press, 1977.

Singer, Peter. Animal Liberation. New York, New York: Avon Books, 1975.

———, ed. In Defense of Animals. New York, New York, Basil Black-well Inc., 1985.

Bailey also will need to work on sentence structure and other writing problems—such as not just preaching at the reader but anticipating the reader's objections and answering them in a persuasive way, and the questions of authorities mentioned in the introduction to this draft—to make his meaning clear and persuasive to his readers. But he has begun the task of collecting information that will influence his readers.

The student begins to understand how to write, and discovers that effective writing, even poetry and fiction, is constructed from accurate, specific, interesting information. When you write a check, it communicates the money you have in the bank, and when you write a draft the words reveal what you know or don't know: the specific information—or lack of information—you have to deliver to the reader.

This is a great comfort to most students and a blow to those students who have been praised for fancy writing in the past. Many students who are not naturally proficient with words become the best writers because they can collect, organize, and order the information readers need. As Ernest Hemingway said, "Prose is architecture not interior decoration."

CHAPTER 6
RE-WRITE TO SHAPE

Now that we know the dominant meaning of the draft and have collected the material from which an effective piece of writing can be built, we are ready to choose the shape or form of the building we will construct.

And you, the writer, need the house of meaning as much as the reader. You have a discovery draft, a rough sketch of your writing, notes and other research materials. Depending on the subject, those notes—mental and physical—may include books and articles you have read, your notes on them, interviews with sources, memories, observations, lab notebook entries, historical documents, reports of experiments, draft fragments of earlier writing, government reports, scholarly monographs—in all, a total mess. Your mental work space looks like a construction site littered with kegs of nails, prefab windows, piles of lumber, shingles, tools, tar paper, concrete steps going nowhere, material filled with potential.

The writer has to control the material, to make sense of it, and does that by shaping the information into a form that gives the material meaning and carries that meaning to the reader. In shaping the material the writer looks to see what connects, what belongs together, what makes sense. Much of the material is interesting, but it doesn't connect with the meaning. It is placed outside the fence of form. The material that does relate and advances the meaning is included within the fence of form. The writer shapes the information and does not so much fill in traditional forms of writing as re-invents the many forms of writing, such as narrative, report, review, argument, letter, proposal.

■ FORM IS MEANING ■

Form gives meaning to your material in somewhat the same way that a house, a barn, an apartment block, a supermarket gives meaning to lumber and nails, steel beams and cement.

In writing, sometimes you can be the architect and design the building of meaning within the limitations of the material and purpose of the building; other times you are the job boss constructing the building you have been assigned from someone else's—perhaps a teacher's—blueprints. In each case, form shapes meaning.

The Unshaped Material

The writing demonstration that follows is constructed from the fact that the writer's parents divorced and his father got married soon afterward. His mother has just remarried the summer before the writer left for college.

Personal Essay

In the beginning I had two parents, then I had three, now I have four. In editing my last paper, the instructor said that "less is more." Well, in remarriage, more is more. Lots more.

I didn't realize when my father moved out how much he would change—and how much our relationship would change. At first he was just angry, then he became like a buddy. He grew young. We hung out on visiting days—at McDonald's, the mall, ball games, watching TV in his small apartment. He told me his troubles—me!—and asked me for advice. And he listened to me and seemed to understand the way he never did when he was my father. Well, he was still my father, sort of.

Mother changed. She became **SUPERPARENT.** She read books and went to meetings and even signed up to coach my soccer team—and she was good. Read books on that too. When she and

Dad were together, she was the easy one. Now she was Ms. Boss—and at work too, where she got promoted. My father grew young; my mom grew old. Not ancient, just responsible, organized. She became an executive mother. She gives me memos on what to do—on office forms!

Then Dad married a woman who made cookies. She made like she was a Disney mother. Annie, she wanted me to use her first name, cooked whole sit down, cloth napkin meals. She gave up her job because she was pregnant. Well, I had stayed over in that small apartment on the living room sofa before they were married. A baby was no surprise to me. It was for them! And I've became her maternity project. It was okay and I like the kid. But it was weird at times, to have a second mother and my not-real one being in a motherly phase.

Now Mom has remarried and I have an extra father. I thought if my mother remarried, I'd have some awkward guy like my friends do, who tries to kid around, be a pal, like my father before he had some else to date. Not this guy. He had six kids before his wife died. He is in charge and I think Mother likes that. He makes charts. We have chores. We eat around a dinner table, all of us. We have topics at dinner and I thought I would hate that but it's fun. He really wants to be a father.

And not only do I have four parents and eight grandparents all living, and three great-grandparents, I, an only child four years ago, have three brothers and four sisters. I cannot count the cousins and the aunts and the uncles. I even have a brother-in-law.

This is a good piece of writing that dramatizes a situation faced by many young people. I am impressed at how compassionate he is and how little he wallows in self-pity. He does not judge his parents and he has a sense of humor. The shape of the piece is clear: a paragraph for each situation in chronological order. The purpose is to look at an autobiographical situation critically. It is a critical essay in the sense that he shows how his parents—all four—change and behave as they take on new roles.

The form doesn't quite work at the end as he lumps in all the other family he has grown in the last few years. This theme could be developed in the first paragraph, or it could be developed into a longer section or even an essay of its own.

Now see what happens as he tells his story in a different form.

Scholarship Application

I am submitting this application to the Hogue Student Foundation because my family's financial situation has changed since. My father has remarried and he has a new baby; my mother has remarried and I suddenly have seven siblings on that side of the family, only one who is old enough to have gone to college. I hope you will consider the changed status of my two families in considering this application for a scholarship.

Letter to a Friend Who Is in the Service

You talk about barracks life. Well, Mother married Mr. Fertile. I have ten people wanting in the bathroom at 7 ayem. I shower at school after practice. They're ok I guess but no privacy. I know what you mean about being alone. I never liked being alone but now I have bunk beds and seven-year-old twins in my bedroom. I can't wait to go back to college and that dorm I complained about last year.

Book Report

In his book, *Role Seeking: The Sociology of Rank-Ordered Adolescents in the Extended Family,* Dr. Finley Robbspierre of the Mescowan Family Research Center, puts the experience of many college freshmen in a social context. At first I felt cheated of my individuality to discover I was part of a familial trend but it was illuminating to find out that . . .

News Report

More than fifty students attended an organizational meeting of Students of Divorce at the Memorial Student Union Monday evening.

Marianne Morison of the University Counseling Service introduced Penelope Stearns-Upton, who is both the daughter of divorced parents and a single mother, and a graduate student doing a doctoral study on the effect of divorce on freshmen.

Her pilot study revealed . . .

Case History

Subject: Myles J. Turner, 17 years old, is a freshman from San Diego, who received word of his parents' separation in the third week of the semester.

Method: The subject was interviewed once a week during the rest of the first semester. The interviews lasted from one to two hours and took place in his dormitory room. They were tape recorded.

His parents and siblings, a sister who is law school in New York City and brother who is married and works as an insurance sales-man, were interviewed by telephone and taped with their permis-sion. All of his teachers this semester, the dorm resident supervisor and his hall resident, his roommate, his girl friend, herself a child of divorce, and three close friends were also interviewed at least twice during the semester and the books and articles on the attached reading list were studied.

Discussion: The subject passed through anger, self-blame, and a beginning of acceptance during this first semester. He was able to articulate the stages through which he was passing and that con-cerned him, "My Dad always had a certain detachment that I didn't like. Every time I screwed up he asked, 'Now what have we learned from this?' I'd like it better if he just got pissed off. I worry I'm stand-ing back from my feelings the same way, that I'm him." . . .

We could go on to imagine a term paper, a letter to a grand-parent who is worried about him, the minutes of a meeting of Stu-dents of Divorce, a personal journal entry, a political science paper on the legal rights of children of divorce, a scene for screen writing class and on and on. Each form would make its own demands as it is designed to achieve a particular purpose and serve a particular reader.

■ DIAGNOSIS: INEFFECTIVE FORM ■

The purpose of all written forms is to carry meaning to a reader. The form must fit the material you have and the audience for that material. The form you choose depends on the expectations of the reader, another way to define literary traditions. Those who give

grants expect a proposal; the literature teacher expects a critical essay; the laboratory assistant a lab report; the history teacher a research paper; the judge a lawyer's brief; the bereaved parent a letter
of sympathy.

Turn yourself into the reader of your draft. Does the material
fit the tradition—the reader's expectation—or break the tradition
and increase its effectiveness? Would you, as a reader, respond
the way the writer expects you to respond? Is this the most effective way to attract and hold the reader's attention?

The ineffective form is one that does not deliver the information the reader needs—the poem sent in as a scholarship application,
the personal letter written in the style of a sociological paper. Scan
your draft, imagining it as a blueprint of the exterior walls of a
building. Does the shape of the building fit its purpose? Is it a factory, a summer home, an apartment house, a supermarket, a medical clinic, a fire station; is it an argument, a narrative, a case history,
a report, a poem, a review, a lab report, a critical essay?

Does your draft, for example, follow the "building code" of an
argument? Does it define and clarify the key issue? Does it make
your position clear? Does it anticipate and respond to the opposi-
tion's arguments? Does it place your own arguments in increasing
order of importance? Is each argument supported with objective
documentation? Is the source of each piece of documentation available to the reader?

Such questions are obvious for each form. Imagine you are the
reader and list the questions that must be answered by the shape of
your draft and the information delivered within that shape. Your
account of an automobile accident can be shaped by the task: filling
out an insurance report, dictating your account to the police, giving a lawyer material for a brief, preparing a eulogy for a victim's
funeral, writing a letter of sympathy to a parent or a personal letter
to a close friend.

In scanning your draft, determine whether the form is appropriate. Does what you have to say fit the tradition, the form the
reader expects?

Many times in school or work, of course, the form is ordered
ahead of the material: it will be a term paper, a corporate memo, an

argument, a case history. Then the focus will be on collecting the material that will satisfy the form.

If the form of the draft is appropriate to the material and to the audience, then you can move ahead to the next stage in the re-writing process. If not, then you may have to choose a different form, turning a report into an argument, an argument into a personal essay.

When the form is required by an instructor or an employer, the writer can only adjust the form—switching the point of view, re-ordering the evidence, casting the traditional form in a manner more appropriate to the message being delivered. Even when you cannot choose the form, you can still make it yours by the way you develop it.

■ SHAPE COMMUNICATES MEANING ■

The shape of what we say contains and therefore helps to communicate our meaning. A house invites us to live in it; a field house invites play, a factory invites productivity. **Story** says there is a beginning, a middle and an end; that action between people shapes events. **Essay** states there is significant information worth critical commentary. A **lyric poem** implies there is truth to be found in image and song. **Description** declares there is something important to describe; **report**, that something has occurred that needs to be reported to a reader; **argument**, that there is something to be argued for or against.

■ DISCOVERING THE FORM OF THE DRAFT ■

To build a house of meaning, a structure that delivers significant information and a critical opinion of that information, you have to chose an effective form. Just as each type of building has a different purpose and a specific structure to serve its purpose, each form of writing is designed for a specific purpose and has a specific structure designed to serve that purpose.

Internal Form

Most writers try to find the form within the draft. They read the draft to see how the information they are collecting dictates the content:

The writer discovers:	the possible form
a significant pattern in ordinary information that the reader needs to understand.	expository essay
significance revealed by a series of events organized chronologically.	a narrative
facts that contradict a law or regulation.	an argument
a situation that needs study.	a grant proposal
a book that readers need to discover.	a book review
an individual who has made important changes in the way we live.	a biographical profile
a personal experience that will give the reader helpful information.	an autobiographical essay
a solution to a problem at work.	a memo

In the cases above, the form may not be clear until the writer has completed a discovery draft. The writer may begin thinking that the material will become an autobiographical essay and it may turn into an argument or a memo; the student may begin to write an argument and discover that the material demands a humorous memoir.

I respect my material and listen to it for what shape it seems to be developing. I think it is important to emphasize the relationship of respect writers should have toward their material. A writer should respect the integrity of the specific pieces of information collected, especially if those details contradict what the writer expected; a writer should respect the patterns into which this material arranges itself—the meaning that arises from the material—especially if it is traitorous to the writer's intent. A writer should respect the message from the language—the music of the evolving text, its voice—especially if does not say what the writer thought it would say or was saying during the writing of the draft.

It is the material, in the best writing, that determines the form.

External Form

Of course, much of our writing must fit an external form that is established by tradition or reader expectation. We are not allowed to look into our material to find an essay, story, poem, history, research paper but must fit our material to an assigned form such as a lab report, critical essay, history term paper.

In these cases we follow the tradition. We manipulate the form we are given so that it delivers a meaning to a reader. We can—and should—be given a model to follow when we are assigned a form that is new to us. If we aren't, we can ask the teacher or the employer for an example or look one up. We may, to our surprise, discover we know more about the form than we realized. That conscious seeking of tradition will connect with the subconscious knowledge of which we are not aware.

Tradition

When facing a writing task you have not attempted before, you can find out what forms have been successful in the past by looking at books that analyze and explain argument, business writing, screen writing, science writing, writing for nursing and police reports, news writing, fiction and poetry writing, scholarly writing, writing criticism and speeches. Such books conduct autopsies of successful writing in that form, revealing the conventions—the tricks of the trade—that have worked in the past.

It is also possible to take a piece of writing you like—or one the audience you are trying to reach trusts and respects—and analyze to see what the author has done. Let me give you an

immediate example of writing and re-writing from my own professional experience.

> Looking back, I realize with horror how tracking affected our lives
> out of school. We hung out with people in our own track. We
> dressed according to track. I could tell a 1 or 4 or 8 or 13 track
> student by how they dressed, the slang they used, the jokes, each
> level's protective snobbery. Our parties often seemed tracked.

This paragraph is from a column I wrote arguing against tracking in school because of the effect it has on students, citing my own
negative experiences with tracking. The paragraph worked and was
published. But in analyzing it, I realize I could put it in a larger
social context.

> When I go back to my home town I am struck at how accurately the
> program that tracked us into college, secretarial, commercial, voca-
> tional and general classes predicted what we would be decades
> later. The program even predicted an amazing number of our dating
> and marriage patterns. I wonder what would have happened had
> we not been tracked. Would the factory worker have become the
> doctor, the doctor the druggist, the engineer the car mechanic,
> the housewife the lawyer, the judge the housewife? Did some test,
> full of racial / gender / ethnic / economic bias, predict our lives be-
> cause we accepted the school's evaluation of our worth as our own?

No magic. Just a careful reading of a text, asking what the
writer is doing, paragraph by paragraph. An effective piece of writing—one that works—will stand up to this type of scrutiny and instruct a writer who needs to understand the form.

────────────── ■ *Writing Exercise* ■ ──────────────

Make a list of what you expect—need—from a piece of writing: a
brochure explaining how to apply for college, a scholarship, a loan;
a story about a championship game; a memo on changing job pro-
cedures; a letter breaking off a relationship; a textbook; an automobile driver's manual. Notice how easily the writer could have made
a similar list. You may want to write this list as questions that demand answers.

Reader Expectation

The experienced writer anticipates the reader's expectations and makes use of them, developing, pacing and voicing the draft to the reader's desires. If the boss wants specific facts, the writer delivers specific facts; if the reader expects understanding, the writer understands; if the reader will only consider an argument that appeals to the brain, the writer serves up thought; if the reader wants emotion, the writer provides emotional material; if the reader is busy, the writer is brief.

The writer becomes the reader, imagines what the reader expects and delivers in many forms of writing. Each form the writer uses has its own pattern of reader expectation, and since we are readers as well as writers, we can predict those expectations.

Writing Against Expectation.

Some of the most effective writing, however, is written against the reader's expectations. The reader expects a sermon and receives a humorous story; the reader expects humor and receives a government report filled with facts that the reader slowly discovers is hilarious; the reader expects an emotional argument and is delivered a list of hard, cold facts; the audience expects a high-flown political speech full of grand clichés and hears a quiet, honest piece of autobiography; the reader of an annual report expects statistics and is taken on a walk through a manufacturing plant.

You can work against reader expectation when the draft communicates significant information in the manner most appropriate to that information; but it is important, in some way, to allow the reader to know that you know the reader's expectations and are contradicting them on purpose.

Or you may be able to accomplish this more subtly, but the reader who comes to the page with an expectation—and all readers do—deserves a response to their expectation. If you don't anticipate their expectation, they will go away early and go away mad, and you aren't an effective writer if your readers leave and don't turn the page.

─────────── ▪ *Writing Exercise* ▪ ───────────
Make a quick list of all the forms your draft might take and then
select the one most appropriate to your meaning and your audience;
then list the reader's expectations from that form and note any way
you might write against those expectations.

─────────────────────────────────

▪ DESIGN YOUR OWN FORM ▪

The writer is the architect of the house of meaning and the writer
works as an architect: selecting from the inventory of traditional
forms; discovering the form that is buried in the drafting board
sketch or draft; or designing a new form to fit this particular
situation.

The Discovered Form

To discover the organic or natural form that lies within a draft, you
can read the draft quickly, trying to visualize the outlines of the
piece, its horizons or boundary fences. There may be several forms
in a draft. It may tell a story and use narrative techniques; try to
persuade the reader and use some of the strategies of argument; relate facts in a manner appropriate to a scientific report. All these
techniques might be used in a magazine article on an environmental problem. But the overall form would be an investigative piece of
magazine journalism. Your job is to discover the form that contains
and communicates the message effectively.

The Invented Form

The invented form is forged from the three principal forces of message, purpose, reader, and once those forces are identified, the
process of invention is usually easy—a matter of simple logic. Usually the inexperienced writer invents—or re-invents—the form after a first draft and the experienced writer before the first draft,
but the process of design is the same.

Message

What is said comes first. The message itself, "I need money," is a force in determining the form. Each message may need a special container that will carry it efficiently to the reader.

Purpose

The purpose of the writer is another force that shapes form. The purpose of the message, "I need money," may be to get sympathy from a friend, to delay a suit from someone to whom you owe money, to make someone pay up who owes you money, or to negotiate a loan from a parent, bank, or college financial aid officer.

Reader

The one who is to receive the message also exerts a profound influence on the form. An appeal that works with a mother may not work with a father. A friend who owes you money may not be influenced by your good grades, but a financial aid officer may be affected by that information.

Create an Effective Design

It is a simple, logical matter to create an effective design, a rhetorical form that is custom-made to communicate a message to a listener and accomplish a clear purpose. Remember, you have been using speech to persuade, report, entertain, communicate since before you started school. You know many of the forms of our literary heritage, you simply don't know the names that scholars use to describe them—and that doesn't matter in the designing process.

Put the following headings across the top of a sheet or pad of paper, even at the top of your computer display screen.

Message ⟶ *Purpose* ⟶ **FORM** ◄——— *Reader*

Write a brief description of your message in the left-hand column; add a statement of purpose in the second column; then jump over *Form* and write a brief description of your reader. *Then* consider the forms that might deliver that message to that reader and accomplish your purpose. It may be the form you have used in the first draft, the form you used with some modifications or a form you have read and written before. It may also be a way—new to you—of delivering a message to a reader.

Here are some examples of how this method might work.

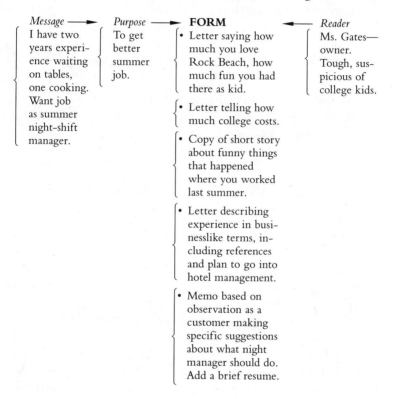

No mystery here. The last one gives the writer the best shot at the job. Try this yourself. The creation of an effective form is a matter of logic.

Using a Traditional Form

The forms you "invent" are usually on the shelf in the rhetorician's warehouse. They may be under such general headings as narrative, argument, report; or they may be categorized by purpose:

to persuade, to tell a story, to explain or to instruct. The forms may also be a specialized warehouse: chemistry lab experiment, sales report, U.S. Army training manual, history term paper, nursing log, engineering field journal.

In whatever field you study, there will be forms that you can try on for size, adapt, and modify to fit your message, purpose, and reader. As you write, you will develop the inventory of forms with which you are familiar, but you should never forget that you can invent the forms you need when you face a new writing task.

──────────────── ■ *Writing Exercise* ■ ────────────────

Use the chart above and design a form that will carry your meaning to your reader effectively.

■ KEEPING IN; CUTTING OUT ■

The effective writer creates from an abundance of information, an abundance of thought, an abundance of feeling. The experienced writer feels the same panic in facing this abundance as the inexperienced writer, but the experienced writer knows that the feeling of going down for the third time under the ocean of information, ideas, and emotions is natural and essential.

Effective writing is the product of a full inventory of material from which the writer can choose to construct a sturdy piece of prose. The writer has to be able to pick the appropriate quotation, the revealing detail, the precise word, the supporting melody, the accurate insight, the fitting feeling, the documenting statistic, from a warehouse of potential material. And, therefore, the revising writer has to master this abundance, making effective use of what is on hand to discover, clarify and communicate meaning.

What Is Saved

All that is saved in the draft is the material that gives the form shape and meaning. Each detail, each word, phrase, line, paragraph must move the meaning forward.

When revising, the writer must, with a cold eye and an icy heart, examine each piece of information the writer has collected, each lovely phrase the writer has created, and save only those that clarify, develop, support, communicate the meaning.

What Is Discarded

Consider Isaac Bashevis Singer's wastebasket. Isaac Bashevis Singer said, *"The main rule of a writer is never to pity your manuscript. If you see something is no good, throw it away and begin again. A lot of writers have failed because they have too much pity. They have already worked so much, they cannot just throw it away. But I say that the wastepaper basket is a writer's best friend. My wastepaper basket is on a steady diet."*

What is thrown into the wastebasket is not wasted. The material not used—that fascinating specific, the unexpected quote, the phrase that once seemed to illuminate, the concept that once appeared to bring all the elements together—led to what was used and what is in the process of revealing meaning, as well as what will be shared in the final draft.

────────── ▪ *Writing Exercise* ▪ ──────────

Make a list, or go through the draft, and mark what must be saved and what can now be let go. Notice how this act of house cleaning simplifies what you have to work on in the revision ahead.

▪ REVISING ACADEMIC WRITING ▪

The challenge of academic writing is to find a way *within* the tradition to express your own critical views, to be individual within formal limitations. To do that you have to first conform to the limitations. In addition to the suggestions made above, many disciplines, through their formal academic organizations or the publications of those organizations, publish guidelines for published research. These can be obtained from a professor, or by writing the organizations. A reference librarian can also show you a directory for such organizations.

Once you are familiar with the writing expected in a field, you can adapt those traditions to the needs of your material. All forms of writing—even academic forms—evolve in response to the latest trends in research. Composition research traditions have changed, as more and more researchers use ethnographic approaches that loom at students in a social context. Most disciplines have changed as computers have become a basic research tool.

In revising an academic paper, it is vital to know the traditions of the particular form your writing must meet.

A PROFESSIONAL WRITER'S CASE HISTORY

Evelynne Kramer, my editor at *The Boston Globe,* asked me to write a piece of personal experience about a summer place or event that was important to me. I immediately thought of writing how I learned to swim, since I almost drowned twice before learning to swim. I saw the piece as a straight, chronological narrative.

> July. Bright sun glints off blue lake water. Watersplash and each drop a bright crystal of light. Rocks rising at the shore where sunladen maple leaves make leaf shadows dance. Dark pine woods, cool and mysterious.
>
> Swimming underwater where the temperature drops and sunlight dims. Flipping over to study the ceiling of water overhead. Joy and terror.
>
> It was in July that I drowned. Twice.
>
> And it was in July that I learned to swim.
>
> The first time I drowned was at Salem Willows. Mother was laughing with some lady friends at the bottom of the pier and I was at the far end, reaching for a row boat when a boy in the next boat, laughing, shoved my boat away with his oar and I fell into the water. I could not swim.
>
> I still dream the crusted pilings racing up, the water turning green then black, the unexpected cold.

I came to on the dock, hands pushing at my back, water spewing from my mouth, and mother running toward me. I did not know mothers could run.

The next time I drowned was at Ocean Park in Maine when a Baptist minister was hired to cure the scaredy-cat little boy of his fear of the water. His was a muscular, male Christianity.

He grabbed me and swam me to the middle of the salt ocean pool where the water was over our heads and tossed me, arcing high in the sky, to where the water was even deeper.

I splashed my way back to him and got him around the neck. I was not holding on. I was determined to kill him for his betrayal—and I almost did.

I remember my thin arms under his chin as I hugged his neck—hard. I remember our going down together, bobbing up, going down again; I remember the glee I felt as I saw his face, close up, go red then dark, anger changing to terror.

He fought me to the edge of the pool where Christians rescued him. I was punished and ridiculed all summer, but I felt enormous satisfaction. I did not go in the water again that July; I did not learn to swim.

And every time I saw him in the pulpit, at the camp, on the sidewalk, I reveled in the look of pure Christian hate and fear he sent me.

It was another summer that I was sent to Camp Morgan, a Worcester YMCA camp in Washington, New Hampshire—a skinny, lonely, only child my uncles felt was spoiled. Camp would make me a man: I would learn to swim.

I was to spend four summers at Camp Morgan, nine weeks each July and August, and my personal geography will always include the Millen Lake, the stumble-rooted trails in the woods, the chapel I helped build, half-mile point, the ball field, Blueberry Hill, a thousand overlapping summer scenes.

I can still feel my legs pushing up Chocorua, the pack high on my back; watch rain create craters in blueberry pancake dough cooking in a frying pan over a smokey, open fire; experience a gentle ride in a canoe and its response to the soft turn of my trailing paddle; discover once more the cellar holes of towns abandoned in the gold rush; wince at the painful surprise of a bee that stung a boy's most private part as he sunbathed on an Asheulot River rock after skinny dipping.

And each July I remember the terror of swimming class, the cruel taunts of instructors and campers, the humiliation of hiding in the woods or crouching in the stinking KYBO [Keep Your Bowels Open] hearing the cheers and laughter, the instructors' whistles and the campers' splashes, and the shouted "Murray" by the counselor assigned to find me and drag me back to the lake.

Shame and ridicule and cruelty at last made me more terrified of not swimming than of water over my head. Day after day, the last boy to swim, I got in the water up to my ankles, then to my knees, then bent over and got my face in the water, at last learned to keep myself moving where I could touch bottom and, weeks later, found myself leaping into water over my head and swimming—but never to this day without fear.

A former teacher, I do not today approve the methods used to teach me to swim. My instructors seemed unnecessarily cruel, gloating at their ability to humiliate. This system did not work for everyone. Some left camp and went home, but camp, even under these conditions, was a happier place for me than home, and I was Scots stubborn. I stuck it out.

After Camp Morgan, paratroop training in a hot southern summer not so many years later was a lark. And when I was in combat I often remembered that July when I was more terrified of retreat than moving further out into water over my head.

Ever since that July sixty years ago when I first drowned, I still fear the terror of water over my head, and when I swim, I slowly move out further and further, my feet touching bottom, then my toes, then nothing.

Once I am in water over my head, my fear under control, I am free to ride the surface of the water, then to dive deep, swimming underwater, slowly turning so I can see the ceiling of water, sunlight refracted by water, remembering the old lessons and remaining calm, not fighting the water, enjoying the beauty of the water world.

It was in November that I had my heart attack, but as they worked over me in intensive care, it was again July, at Camp Morgan, and I repeated the lessons I learned there: to survive, accept this changing world, do not waste your strength in struggle and splash; focus on staying afloat.

The nurses congratulated me on my attitude, not knowing I was once again learning to swim. Following the lessons of July, I made it to shore again, and in the unexpected life that follows I also remember the July lesson to enjoy the beauty of the world.

The terror of drowning—of aging, of leaving this summer world—is still with me, but it is under control. And the terror makes the fragile beauty of water spray against the sun, the pattern of leaf shadow, the mystery of dark pine woods, more beautiful each July.

After lunch Evelynne called to say she has reservations about the July learning-to-swim feature. I offered to dump it; but she said it could run, though she thought it could be better.

She thought it took a long time to get into it. I suggested that place was at paragraph 17. She thought it was further down, *four* paragraphs from the end!!!

More notes I made while she was on the phone:

- gets good as it gets rolling
- long time to get to meaning
- lot of time spent introducing characters not significant to me
- perhaps rethink the narrative
- get to the point about death, fear, heavy duty stuff

I was fascinated by her comment that the piece started four paragraphs from the bottom, and that starting at that point would change the piece from a narrative into an essay reflecting on a series of experiences. I started at the end, re-writing by working within the draft to see what this new form would tell me.

The following draft is shorter, tighter, more immediate.

You may find it interesting and instructive to take your draft as I am going to take mine, and go through it paragraph-by-paragraph to see just what I have done in the revision. It can also be helpful to do this to a piece of writing you like.

It is again July, and I stand in water up to my chest. I try to talk away the fear of my drownings, fail and suddenly make myself spring up and then down, swimming underwater, out where my feet cannot touch bottom.

REMEMBER: *What I say in this commentary is spoken by the editing me who must be rational and objective about what the writer me did instinctively. Now I stand at a distance and rationalize what I did when I was unconsciously instructed by the draft and my craft.*

I want to put the reader into my shoes—oops, bare feet—so the reader can experience my fear and my response to it. I also want to establish the tension that will drive the piece: my fear of swimming and my delight in swimming.

I hope for reader identification, that readers will begin to experience their own July's as they read mine.

And a hint of something mysterious—"my drownings"— that may make the reader read on. Also an echo to that moment of terror we all felt when our "feet cannot touch bottom."

And all in 42 words.

The water is cold, dark as I remember, and then I swim up to light, break the surface and tread water, my fear under control until another summer.

I get the reader in the water.

July is always a month of joy and terror, the month I twice drowned, the month I learned to swim.

This is the context paragraph, the place where the personal experience essay to be told is placed in a larger focus. The reader is promised, by implication, that the individual experience will have a significance that goes beyond its importance to the narrator.

I develop the tension, working from specific, physical details to more psychological issues. I look back to the paragraph, fourth from the end, of my first draft, to see how I took off from there. In a way, writing is like jazz improvisation, picking up a theme and, within a tradition, running with it.

Now I am free once more to enter the water life—pool, river, pond, ocean. I ride the surface, jackknife and dive deep again, swim underwater—my greatest joy—slowly turning so I can see the ceiling of water, sunlight refracted by water, remembering the old lessons and remaining calm, not fighting the water, enjoying the beauty of the water world.

I want to share my celebration of swimming with the reader. I did not realize the importance of "remembering the old lessons and remaining calm, not fighting the water" when I first wrote it, but it now foreshadows what will come in the piece and will be repeated in slightly different words in the third paragraph from the end. Those words contain the ultimate meaning of my essay; but I did not know that when I wrote them here and I did not remember these words when I wrote them at the end of the piece. I feel as if this meaning was discovered and woven through the piece by the text. My job is to read and follow the evolving text—the evolving thinking that is writing.

I suppose the terror of drowning—of aging, of leaving this summer world—makes the fragile beauty of water spray against the sun, the pattern of leaf shadow, the mystery of dark pine woods, more beautiful each July.

Now I echo back to my first "poetic" beginning as well as to the terror and deepen the feeling in the piece that is, in a way, about dying. I see that now, but didn't know it while writing drafts. The themes of writing come from the writing, rarely before. Once more I am aware that writing is thinking, not previous thinking written out.

The first time I drowned was at Salem Willows; I was eight years old. Mother was laughing with some lady friends at the foot of the pier, and I was at the far end, reaching for a row boat when a boy in the next boat, laughing, shoved my boat away with his oar and I fell into the water. I could not swim.

Now I am back to the chronology but I move faster than I did before. I have a complicated series of actions to make clear. To do that, I see the scene with memory's eye and record what I see—in sequence.

The point of greatest impact or emphasis comes at the end of a paragraph—or a sentence or a piece of writing. Short sentences also give emphasis. Note my short sentence at the end of this paragraph.

I still dream the crusted pilings racing up, the water turning green then black, the unexpected cold.

I want to put the reader in the water, drowning. Note the action: what's going up fast means I'm going down fast; the color of the water, then its temperature.

I came to on the dock, hands pushing at my back, water spewing from my mouth, and mother running toward me. I did not know mothers could run.

A touch of humor at the end of the paragraph to lighten the seriousness of the experience. Of course I didn't think that and put it in. And writing this I realize the humor is black. I was surprised that my mother was so concerned about me—perhaps a cruel and unfair thought but a true one. I wrote it because that was what I remembered as I re-created the experience in my mind and on paper. In re-writing, I cut, save or develop, rationalizing what I wrote instinctively.

Each July, when I dive down into the cold water below the surface, I see those pilings rushing up.

I connect past and present.

The next time I drowned was at Ocean Park in Maine when a Baptist minister was hired, a couple of summers later, to cure the scaredy-cat little boy of his fear of the water. Ours was a muscular, male Christianity.

In writing the first draft, I discovered I am still, perhaps even more, angry with that minister now than I was when I was a boy and subjected to the behavior of grown-ups. John Hawkes says, "Fiction is an act of revenge." Perhaps, in part, all writing is an act of revenge, a very uncomfortable thought.

He grabbed me and swam me to the middle of the salt ocean pool where the water was over our heads, and tossed me, arcing high in the sky, to where the water was even deeper.

I hope the reader is in the air with me.

I splashed my way back to him and got him around the neck. I was not holding on. I was determined to kill him for his betrayal—and I almost did.

I remember my thin arms under his chin as I hugged his neck—hard. I remember our going down together, bobbing up, going down again; I remember the glee I felt as I saw his face, close up, go red then dark, anger changing to terror.

He fought me to the edge of the pool, where Christians rescued him.

A series of very complicated actions to make clear and some not so complicated emotions to reveal. Revenge? Well, I guess so. Glee—just the right word. I cut out the external comments about teaching philosophy and methodology from the previous draft. Show and there may be no need to tell.

He taught me what I did not know: there was a man inside of me, a tough, cold man of rage that I met years later when I served in combat as a paratrooper.

But I do need to put the experience in a larger context right now. What did it mean to me—and perhaps to my readers? What did I learn and what might my readers learn? Too often personal experience pieces remain in the personal; their significance is not revealed. This is not just the story of a spiteful kid, no matter how justified. It did mark me in a way I—and the reader—do not expect but understand once it is said.

When I finally learned to swim at Camp Morgan—the Worcester YMCA Camp then at Washington, N. H.—the mechanics of the crawl, breathing to the side, the overarm stroke, the flutter kick were easy. It was the fear that was hard.

More context and the specific details give it authority and make the writing lively. We write with specific information.

At first I would not go to swimming class, finding my fear of water greater than the taunts of instructors and campers, the humiliation of hiding in the woods or crouching in the stinking KYBO [Keep Your Bowels Open] as the counselor assigned to hunt me down shouted "Murray" until he found me and dragged me to the lake.

Now I can begin to create the climate of counter-fear, of ridicule and shame, that made it worse for me not to swim than to swim.

At last I learned to jam the fear down inside me, to control the shaking, the tears, the cramps in my gut, and to dare the water, at first up to my ankles, then my knees, at last, weeks later, over my head.

On this reading, I think I hear the reader asking, "How did you control the fear?" I stop right now and create a new paragraph. But when Evelynne edited the story she had problems with the paragraph that piled up like a train wreck. I agreed. Too much, too much.

I quickly rewrote the paragraph three times, making it tighter, cleaner. Evelynne liked the result.

(Original)   I learned how we control fear intellectually, observing that others can swim, others that are often more inept that I am; I learned how we control fear physically by taking deep breaths, forcing our muscles to relax; I learned how we control fear socially by being more afraid of not doing what others do than of doing it; and I learned, most of all, to abandon myself to terror, giving up hope and diving in.

(1st rewrite)   I learned how we control fear intellectually, observing that [added] others who had knees that drummed against each other, lips that trembled yet moved further and further out into the lake **[cut]** ~~still can swim, others that are often more inept that I am~~; I learned how we control fear physically by taking deep breaths, forcing our muscles to relax; I learned how we control fear socially by being more afraid of not doing what others do than of doing it; and I learned, most of all, to abandon myself to terror, giving up hope and diving in.

(2nd rewrite)   I learned how to control fear by mobilizing the fear of not measuring up to counterattack the original fear, taking a deep breath, abandoning myself to terror and diving in.

(3rd rewrite)   I learned how to mobilize the fear of not doing against the original fear of doing and, in that moment of calm, take a deep breath, abandon all hope and dive in.

Now this crucial paragraph the reader demands is clear as I use small words in a direct manner. I worried about the echo of the reference, "Abandon hope, all ye . ." and decided it would work for people who did not get the reference and enrich the text for those who did.

After Camp Morgan, paratroop training in a hot southern summer not so many years later was easy—well, relatively easy. And when I was in combat, I often remembered the July when I had to dive into water over my head.

This continues to reveal the significance of what I learned at summer camp. I also hope the reader will feel again the childhood terror of water over your head so they will accept the larger terror that follows. Now I am coming to the end—and the full meaning of the piece.

In the November I had my heart attack I silently repeated the lessons I learned at Camp Morgan: Do not waste your strength in struggle and splash, but accept the world you fear; relax and float on the unknown.

What I learned is spelled out in a still larger, more immediate context. The reader may apply these lessons to the reader's own experience.
Remember what I wrote in the fourth paragraph.

During the attack and afterwards the nurses congratulated me on my attitude, not knowing I was once again learning to swim.

Busy, busy, tying all the threads of the essay together.

Each July in the unexpected life that followed, I stand in water, then force myself to jackknife up, then dive down, turn, rise easily to the surface, float on what I fear, grateful for yet another July.

"Float on what I fear." Yeah! Go get 'em Murray.
And a return to the joy of swimming now experienced in depth—pun intended. The essay on learning to swim becomes an essay on learning to live while dying as we all are. The reader will not articulate that but many will feel it and be moved by it as they re-live their own living in the face of death.
Did I know it meant that? No. Not when I wrote it. Not when I revised and edited it. Not when I read it in the paper, but only now after writing this commentary. The writer, after

all, is the first and last reader; the writer writes and re-writes to discover what the writer has to say. The surprise of what I have to say that I did not know I had to say is what draws me to my writing and re-writing desk every morning. I thought I was writing about learning to swim.

Shaping becomes one of the writer's great delights. We all write against chaos, trying to understand the confusion of life. By leaving some material out, including other material, ordering it, discovering pattern, we begin to see meaning. As we continue to shape the material during the process of revision, meaning becomes increasingly clear—to ourselves and to our readers.

CHAPTER 7
RE-WRITE TO ORDER

Readers should be drawn into a piece of writing, so they will follow a trail that leads to meaning. The writer must create a path of continual seduction that keeps readers interested and eventually satisfy them.

Readers are in control. Readers can leave the text at any time and will if they don't have a sense of progress. They may not be told explicitly where the reading is taking them, but they need to have a sense that they are moving toward meaning.

Re-writing effectively means reading what you have written with a stranger's eyes to see if there is a clear, seductive trail of exploration that runs through the draft. That line may be clearly marked by headings or it may be hidden within the material, but readers must sense a sequential order in what they are reading or they will put the writing down.

▪ DIAGNOSIS: DISORDER ▪

In diagnosing a draft, there are many signs of disorder, the lack of a sequence of information that allows the reader to accumulate sequential information that becomes meaningful. Some of the signs of an ineffective order are discussed on pages 126-128.

1. The trail through the draft is based on a sequence of false assumptions or assumptions that are true only to you, the writer. Count the false assumptions in the example below. I count three whoppers.

Example:

> We should not send good tax dollars to foreign countries. Other countries do not have the moral standards of the U.S.A. Better to use those bucks at home, in our cities, for example, where people cheat welfare but at least it is on the economy like at least if you make a bad loan it should be to someone in the family, then there's no trouble.

2. The writer tells us the emotions the writer wants us to feel, but we need less emotional direction and more details that make us feel for both people. The reader should have the emotions; the writer should inspire those emotions from what is written on the page.

Example:

> I was so overwhelmed when I saw my father, after all this time, and in a hospital bed. It was so sad, I almost cried. In fact, I did cry, the hospital bed and everything. I remembered him young and he was so suddenly old. All the feelings of hatred and resentment and anger and loss surged up within me. And yet he was my father. It was tragic, so sad for him and for me.

3. The draft was written by a kangaroo who takes great leaps for reasons the reader does not understand.

Example:

> If we are going to do something about the deficit we should start by paying ball players less. The owners of professional teams make money—so do universities from football—but the workers earn more than is appropriate in a nation that has a foreign trade imbalance and a national debt to reduce. Taxes on sport tickets might help.

4. The writer is lost in the accumulation of information that seems to have no context, no fascination for anyone but the writer, no order.

Example:

> I have put a new 486 motherboard in my 55sx but I don't need to in my laptop, new, that already has a 486, both are 25mg, with lotsa hard ram and expanded RAM of course but only one has a built-in fax/modem. The other is external, like my tape back-up.

5. The writer wanders off the trail to examine wild flowers, butterflies and mountain streams that are interesting but have nothing to do with the subject at hand.

Example:

> Trust is the most important element in a relationship. I certainly discovered that after the accident in the sculpture studio last week when my roommate trusted her partner and lost three fingers to a power saw a classmate said he knew how to operate. I wonder the role of trust when people perform dangerous jobs in the military, as lumberjacks or as farm worker, especially children.

6. The re-writer asks "How does this piece of information advance the reader's understanding of the subject?" and finds there is no answer.

Example:

> Automobile manufacturers have adapted vehicles to the needs of their customers and to the dreams of their customers—they may live in an urban area but they like to drive adventure vehicles, four-wheel drive and all, as if they were pioneers. These vehicles allow the driver to sit up high and they are ready to ford rivers, up canyon arroyos, whatever, drive through snow drifts in the sunbelt.

7. The sequence is unnatural, the reader has to leap back and forth in time or logic for no artistic reason going from A to F to C to G to K to B.

Example:

> There are things I know now about college sports recruiting that I didn't when I needed to. I can't blame my parents. They never went to college, to high school. What my high school coach told me was all wrong. I can't believe what he said. And it wasn't just this school. I was made promises like you can't imagine. And no-one told me what would happen to me if what happened happened. Now I know and it's too late. But maybe the appeal will help.

8. The draft reads like Swiss cheese, it is full of holes.

Example:

> The latest treatment for diabetes, at least in tests, is far better than current practice and will be appropriate for all kinds of diabetes although the treatment will differ unless new technology is developed, as has been announced, and its cost is cheaper than what patients have to pay now, a substantial figure at best.

9. The reader's questions that we discuss in the next section are not answered the moment they are asked. The effective writer can predict when the reader needs a definition, documentation, an example, more description by hearing the reader's subconscious questions: What's that word mean? How come? Who says? So what?

Example:

> "Wysiwyg" has made it much easier for me to write textbooks on the computer, of course that has always been true for Macs, but I have had to live in the Dos world and that has limited me until I moved to WordPerfect® 6.0 and chose graphics mode.

■ ANSWER THE READER'S QUESTIONS ■

I have found one organizational technique more effective than any other in revision: anticipate and answer the reader's questions.

Any piece of writing is a conversation with a reader who interrupts to say:

"How come?"

"How do you know that?"

"Says who?"

"I don't get it."

"What do you mean?"

"I'd like to know more about *that.* "

"No kidding."

"Why'd she do that?"

"What'd he do then?"

"Tell me more."

"Stop it. Enough already."

"Get to the point."

"Whoa. Back up, I don't understand."

"Whatta you mean 'gaseous diffusion'?"

Inexperienced writers—and some experienced ones—do not hear that half of the conversation. All effective writers hear the reader's questions and *answer them the moment they are asked.*

Write down the questions the reader will ask about your draft:

- Who is most responsible for the passage of Civil Rights Act of 1964?
- Who was for it? Against it? Why?
- What were the federal civil rights laws— if any—in place before the Civil Rights Act of 1964?
- What did it accomplish?
- Why was it needed?
- What did the law say?

I usually find that there are four to six questions that *must* be answered to satisfy the reader. They are the simple, obvious questions that someone who is deeply involved in the subject may forget, but the common reader will ask them. They may even be questions the writer does not want to have asked, but there is no escape—those questions will be asked.

——————————————— ■ *Writing Exercise* ■ ———————————

Read a piece of published writing you like. Write down the questions that come to your mind as you read—what you need to know to read on—and then describe, after each question, how the writer has answered, failed to answer, or answered the questions in the wrong place when it is too early or too late.

After the re-writer has determined the questions, the order in which they will be asked can be anticipated. The question that is third from the end in my example should be last and the one second from the end should be first (and the third question can be incorporated into it). Rearrange the questions in the order the reader will ask them and re-write them if necessary:

- Why was the Civil Rights Act of 1964 necessary; weren't there laws against discrimination?
- Who is most responsible for the passage of the Civil Rights Act of 1964?
- Who was for it? Against it? Why?
- What did the Act provide?
- What did it accomplish?

Can you use this technique *before* you write the first draft? Of course—and I do—but I still have to use it in revising and in editing other people's copy. A version of this that I have used in revision is to write the reader's question, which the writer is answering, in the margin of the draft and then re-order the piece to anticipate the sequence in which the reader would ask them.

■ OUTLINE AFTER WRITING ■

Outlining has been emphasized as a re-writing or planning activity, and of course it can be. But to do an effective outline in advance of writing, the writer has to have a pretty firm idea of what the draft is going to say. And many times, since writing is thinking, it is impossible to write an outline ahead of time, or if it is written ahead of time it bears little resemblance to the first draft.

But it is always helpful to outline during revision to reveal the structure of a draft, and then to design the structure that the draft must have to satisfy the reader.

To Expose the Structure of a Draft

Some of the ways the re-writer can strip away the language from the draft and reveal the structure underneath are:

- Ask the reader's questions as described above.
- Read through the draft and make a formal outline to visualize the structure of the draft.
- Use the movie writer's storyboard and put each topic you come to on a slip of paper with a reference to the page and line in the draft—statistics on rural poverty p. 3, 1.13–22. Then rearrange the slips of paper into a logical sequence.
- Create a computer tree or draw one that shows the sequence of major points in the draft.
- Draw a map or graph that shows how the major issues rise, fall and interact with other issues in the draft.
- Write a quick, shopping-list sketch of the main points in the draft.
- Make a computer printout of the draft on your screen; underline or otherwise mark the key phrases; then cut away the rest of the draft to see the structure revealed.
- Write out the questions asked by each section and then look to see if they are answered in the following section.

- Write down the major section headings to see the line—the logical order—that is the skeleton of the draft.

Adapt the Structure

Once you see the structure of the draft, you can often imagine the structure the reader needs. It may be easy to adapt your draft to the needs of the reader by moving sections around, perhaps creating new ones and eliminating old ones.

Such moves may be suggested by test readers—classmates, workshop members, instructors, editors, friends, family—who, because of their need to understand and their distance from the writing of the draft, see a potential new structure clearly.

Re-Design the Structure

Many times the structure of the draft has to be abandoned; it just doesn't do the job of producing a draft that can be understood by the reader. In this case you have to design a new trail through the material.

This is a good time to work backwards:

- Write down at the bottom of a page what you want the reader to think and feel after reading your draft.
- Pick a starting point in the material that is as close to the end as possible while including all the information the reader needs to arrive at the conclusion you have written down at the bottom of the page.
- Note the three to five pieces of information the reader needs, in sequence, to arrive at your ending.

Of course you can design an outline with the reader's questions or with the other forms of outlining described on pages 104-112.

──────────────── ▪ *Writing Exercise* ▪ ────────────────

Outline a draft you have recently written on the left-hand side of a piece of paper; then, on the right-hand side, see if you can outline it so that it will better serve the reader.

▪ REVISING ACADEMIC WRITING ▪

Order is of enormous importance in academic writing, which is always concerned with critical thought and an orderly intellectual movement from point to point.

Some disciplines have formal orders that should be followed, unless it is necessary to design a new order to communicate your meaning. And, to be honest, few academicians will allow beginners to do that. Find out if your instructor or editor has a structure you must follow in writing a laboratory experiment, a business or sociological case history, a book review.

STUDENT CASE HISTORY

Silke Burnell

▪══════▪

Silke Burnell, a native of Germany who came to the United States when she married a U.S. citizen and has been working in nursing homes in both countries, explains why she chose her topic: "The idea for my argument paper resulted from my being annoyed about the care provided in American nursing homes and feeling sorry for the residents." She continues, "Because of my six years of work experience in German nursing homes and having worked in American nursing facilities for a year, I could compare the level of care given at nursing homes in both countries. The paper began when I listed all the details I wanted to discuss. Most of the facts in the draft came from my work experiences, but to support what I already knew, I gathered some reports from magazines and newspapers at the library. Having my thoughts down on paper made it easier for me to understand why I had such a hard time adjusting to the care provided here."

In the following revision to thinking-aloud writing, Burnell establishes the points she needs to develop to compare U.S. and German nursing homes and offer possible solutions to U.S. problems. "Through revising my first draft," Burnell explains, "I realized that I should explain things in more detail, because the reader wouldn't have enough knowledge about the daily routine in both German and American nursing homes. I also learned that it would be more effective to generalize my personal experiences to keep the reader on the subject."

Her first draft:

A Change Is Needed

It's strange that in this society, which appears to be so advanced, elderly care is still in its infancy. Apparently the level of care in nursing homes doesn't go past the basics. During conversations with nursing home employees, I was surprised to learn that they had no idea that there was so much more to provide. People must be made to understand that there is a problem, and a new type of profession should be created (and respected by the general public)— Altenpfleger ("geriatric care giver")

When I look at the nursing home situation around here, I feel like I've gone back in time. An occupation that is happy to restrain "only 19%" of its patients, compared to 40% three years ago (Belkin A24), severely needs to catch up to other industrialized countries. I have seen nurses administer tranquilizers to residents that were only reacting to their restrained, shut-away existence. Sedating boisterous residents is no substitute for meaningful activities and stimulating conversations.

The bureaucracy involved with medications, and the time spent administering them, takes up about half of a nurse's working hours. The remainder of their time is spent documenting residents' physical and mental condition, even if they don't provide patient care. Six years experience as a night nurse in German nursing homes shows me that this could be done differently. In comparison, it took me only an hour to prepare an entire day's worth of medication; Altenpfleger on the day staff spent only a few minutes handing them out. That left the rest of the time for patient care and rehabilitation.

Another factor that limits performance is the overabundance of supervisors. A recent study showed that, nationwide, there is one supervisor for every seven full-time staff members (Sheridan, White and Fairchild 335). I have worked in nursing homes where the ratio is as high as 1:2, and the supervisors have only limited contact with the residents. This results in each caregiver having to provide total care to at least 8 residents within two hours, depending on which shift the person is working. In Germany, the ratio is closer to 1:12, and the supervisor also provides full patient care when necessary. Reducing the number of supervisors here, and, or having them work more with the residents, would improve our quality of care.

Nursing home quality would benefit most from having specialized people providing care. Instead of that, facilities cut costs by hiring certified nursing assistants (CNAs). This vast majority of people working in nursing homes have only limited training, a two-week (80-hour [Bowers and Becker 360]) crash course in how to perform patient related care. They learn only the mechanics of cleaning and feeding, not the psychological and emotional consequences for the residents. Because of this lack of knowledge, and the limited time available, some CNAs form a "maintenance opinion."

> . . . that elderly residents should be discouraged from being involved in social and personal care activities, that they are difficult to understand, and that they need to be closely monitored (Sheridan, White and Fairchild 337).

Every minute counts, and caregivers learn to see residents as no more than parts on an assembly line. There is no time left for personal interactions. Immobile patients do not get the chance to take part in social activities, because the majority of employees in nursing facilities lack the understanding that a brain needs stimulation to continue to perform.

In Germany, becoming an Altenpfleger takes at least a year and a half of combined theoretical and practical education, along with passing a difficult exam. Not only do Altenpfleger learn how to clean, feed and medicate residents, they are also taught how to meet the psychological and emotional needs of the elderly. They are also involved in rehabilitating residents, which eliminates the need for a separate profession of physical therapists like you find over here.

Caring for the elderly is an important job in society, but the CNAs that perform most of the work are not respected. In the hierarchy of the nursing home, the CNAs have the lowest value. They deal with the residents most, but they have no voice in discussions about them. The nurses meet regularly to discuss care plans, but exclude CNAs who know the most about the patients. If a shortage of CNAs occurs, those that have already worked are expected to stay another shift, or those with a needed day off are called, instead of the nurses filling in. Those who cannot work a second consecutive shift are looked down upon. Considering all these factors, it isn't hard to understand an annual CNA turnover rate of 120% to 145% (Bowers and Becker 361).

There are many ways to improve CNA working conditions, and therefore the quality of care. The division of workload could be adjusted between the shifts. For example, the night shift could take on some responsibility for getting restless people out of bed and providing personal care. This would reduce the very heavy effort of the morning shift. Also, the requirement to document almost everything the residents do could be relaxed, leaving more time for personal interactions. The daily schedule could be altered to leave more space between meals. Most importantly, teamwork should be encouraged instead of criticized.

I was trying to achieve these goals when I was employed at a nursing home, attempting to speak to the nursing manager. She showed apparent interest at the beginning, but never got back to me to discuss any details. Whenever I talked about these ideas with my co-workers, I got a negative response, since they didn't want to let go of what they were used to.

Hopefully, in the near future, an open-minded person will see the need for changes to bring the U.S. in line with other industrialized countries, when it comes to elderly care. If this doesn't happen, nursing homes will become more expensive, but the quality of care won't improve. And our parents' generation or we ourselves will pay the price.

The order of the material in the paper is effective, but she needed to develop parts of the paper. Remember, it is just as important to leave in what works as to fix what doesn't work. To prepare for her revision of the final draft, Burnell wrote, at the instructor's suggestion, four questions that the reader would ask:

- Why is it important to provide a homelike atmosphere?
- How does the Altenpfleger know if there is an emergency in one of the apartments?
- Why are they (the Altenpfleger) taught not to restrain residents?
- What are the supervisors doing all day long?

Then Burnell answered them in her final draft. I have not made special note of her editing but have put in boldface the large chunks of revision with which she developed her argument:

A Change Is Needed

It's strange that in this society, which appears to be so advanced, care for the elderly is still in its infancy. Apparently the level of care in nursing homes doesn't go past the basics. During conversations with nursing home employees, it is surprising to learn they have no idea that there is so much more to provide. People must be made to understand that there is a problem, and a new type of profession should be created (and respected by the general public)—Altenpfleger ("geriatric caregiver")

The importance of providing a homelike environment for residents of nursing homes has long been recognized in Germany. Entering a nursing home there gives you the feeling of visiting someone at home. Each resident or couple has a one- or two-room apartment, complete with bathroom and kitchenette. They also have complete privacy, with their own apartment key. Altenpfleger are allowed to intrude on this privacy only in case of an emergency. Residents aren't moved from their rooms for any reason. If they

eventually need intensive care, they remain in their familiar surroundings.

At mealtime, each person has the choice of eating in the dining room or having room service. Dependent residents are encouraged to eat in the dining room with all the other people, to ensure they don't feel left out. An attendant is always available in case they need assistance. Another important factor in the daily routine is activities, such as physical therapy, arts and crafts, music, dancing and games.

All these activities are provided by Altenpfleger. They undergo a two-year education, which can be either vocational or co-op in nature. Their broad schooling, which focuses on the psychological, emotional and physical decline of the elderly, is followed by a difficult exam. Both during and after their education, Altenpfleger are qualified to perform duties such as personal hygiene, changing dressings, determining and administering medications (also in emergencies), and calling a doctor or ambulance when necessary. They're taught to avoid physically restraining the residents; instead, they encourage confused people to take part in activities.

When I look at the nursing home situation in the United States, I feel like I've gone back in time. An occupation that is happy to restrain "only 19%" of its patients, compared to 40% three years ago (Belkin A24), severely needs to catch up to other industrialized countries. I have seen nurses administer tranquilizers to residents that were only reacting to their restrained, shut-away existence. Sedating boisterous residents is no substitute for meaningful activities and stimulating conversations.

Here, the bureaucracy involved with medications, and the time spent administering them, takes up about half of a nurse's working hours. **He/she must push a cart from room to room, check the cart and prepare each dose separately. After administering each dose, the nurse must sign the chart to insure that the resident received the medication.** The remainder of their time is spent documenting residents' physical and mental condition, even though they depend on certified nursing assistants' (CNA) reports for their data.

In comparison, a night nurse in a German nursing home spends only a small portion of her shift preparing for the next day's entire schedule of medications. There are 3 trays, one each for the medication to be given after breakfast, lunch and dinner. Each patient has a chart with a medication assignment, where you can clearly see which dose has to be administered, and at what time. The night nurse puts the medication together, and inserts each dose in a cup. These cups go into small compartments in the trays, and each compartment is labeled with the resident's name and required dose. One Altenpfleger of the day staff spends only a few minutes, total, handing the medications out. This efficient procedure gives the staff more time for patient care and rehabilitation.

Another factor that limits performance here is the overabundance of supervisors. A recent study showed that, nationwide, there is one supervisor for every seven full-time staff members (Sheridan, White and Fairchild 335). I have worked in nursing homes where the ratio is as high as 1:2, and the supervisors have only limited contact with the residents. This results in each caregiver having to provide total care to at least 8 residents within two hours, depending on which shift the person is working. In Germany, the ratio is closer to 1:12, and the supervisor also provides full patient care when necessary. Reducing the number of supervisors here, and/or having them work more with the residents, would improve our quality of care.

Nursing home quality in the United States would benefit most from having specialized people providing care. Instead of that, facilities cut costs by hiring certified nursing assistants (CNAs). This vast majority of people working in nursing homes have only limited training, a two-week (80-hour [Bowers and Becker 360]) crash course in how to perform patient related care. They learn only the mechanics of cleaning and feeding, not the psychological and emotional consequences for the residents. Because of this lack of knowledge, and the limited time available, some CNAs form a "maintenance opinion."

> . . . that elderly residents should be discouraged from being involved in social and personal care activities, that they are difficult to understand, and that they need to be closely monitored (Sheridan, White and Fairchild 337).

Every minute counts, and caregivers learn to see residents as no more than parts on an assembly line. There is no time left for personal interactions. Immobile patients do not get the chance to take part in social activities, because the majority of employees in nursing facilities lack the understanding that a brain needs stimulation to continue to perform.

Caring for the elderly is an important job in society, but the CNAs that perform most of the work are not respected. In the hierarchy of the nursing home, the CNAs have the lowest value. They deal with the residents the most, but they have no voice in discussions about them. The nurses meet regularly to discuss care plans, but exclude CNAs, who know the most about the patients. If a shortage of CNAs occurs, those that have already worked are expected to stay another shift, or those with a needed day off are called, instead of the nurses filling in. Those who can not work a second consecutive shift are looked down upon. Considering all these factors, it isn't hard to understand an annual CNA turnover rate of 120% to 145% (Bowers and Becker 361).

There are many ways to improve CNA working conditions, and therefore the quality of care. The division of workload could be adjusted between the shifts. For example, the night shift could take on some responsibility for getting restless people out of bed and providing personal care. This would reduce the very heavy effort of the morning shift. Also, the requirement to document almost everything the residents do could be relaxed, leaving more time for personal interactions. The daily schedule could be altered to leave more space between meals. Most importantly, teamwork should be encouraged instead of criticized.

Trying to achieve these goals while employed at a nursing home is a frustrating process. Nursing superiors show apparent interest in changes, but never follow through to discuss details. Talking about these ideas with coworkers garners a negative response, since they don't want to let go of what they're used to.

Hopefully, in the near future, an open-minded person will see the need for changes to bring the U.S. in line with other industrialized countries, when it comes to care for the elderly. If this doesn't happen, nursing homes will become more expensive, but the quality of care won't improve. And our parents' generation, or we ourselves, will pay the price.

Works Cited

Belkin, Lisa. "Nursing Homes Try Life without Restraints." The New York Times 24 Mar. 1993: A24.

Bowers, Barbara, and James Becker. "Nursing Aides in Nursing Homes: The Relationship between Organization and Quality." The Gerontologist 32.3 (1992): 360–61.

Sheridan, Laurie, Kelly White, and Cathlynn Fairchild. "Ineffective Staff, Ineffective Supervision, or Ineffective Administration? Why Some Nursing Homes Fail to Provide Adequate Care." The Gerontologist 32.3 (1992): 335–37.

Because the shape of the writing and the order of the material within it is clear, I feel I have turned the last corner and can see the finish line. I head down the home stretch. All the most important problems are solved; the rest is craft.

I delight in craft, the fitting in of material that will satisfy the reader's hunger for evidence and information. I revel in the joy of language, when I erase all evidence of revision and craft, making the final copy look spontaneous, easy, clear, and I hope, graceful.

CHAPTER 8
RE-WRITE TO DEVELOP

The most critical difference between poor, unread writing and fine, well-read writing is development. People imagine that a creative idea—the more eccentric, the more creative—is what marks good writing. But there are few if any new ideas; the creativity and the quality comes in the development of a piece of writing.

Fine writing makes the writer's vision of an idea, a place, a person, an event clear to the reader with a rich blend of revealing, specific details, observations, references, patterns of thought—all the forms of information appropriate to the subject.

It is also the richness and fullness of the text that gives the writer authority and makes the reader trust the writer.

Joseph Conrad said, "My task . . . is, by the power of the written word, to make you hear, to make you feel—it is, before all, to make you see." The writer recreates experience—with texture and in depth—so that the reader sees, feels, thinks.

■ DIAGNOSIS: SUPERFICIAL ■

Undeveloped writing is difficult for the writer to diagnose because the draft is fully developed in the writer's mind; it just isn't on paper.

The writer writes *"It was terrible accident"* and sees the victims trapped in the car, hears their cries for help, hears the sirens, is bathed in the flashing light, sees the rescue workers using the jaws of life, sees the blood, the body rushed to the ambulance, bottles with intravenous tubes held high, sees the next victim wheeled slowly to the next ambulance, a blanket over the head. The writer has only delivered a blank check, *"terrible,"* to the reader.

Another writer says it is *"a good idea to raise gasoline taxes"* but the reader asks, "Why?" and finds no answer in the text.

The readers of these two types of writing will not read on. For this reason, the writer has to develop the ability to stand back and create a distance between herself and the draft so that she can read the draft as a reader—hungry for information—will read it.

The signs of an undeveloped text are many: it is predictable; it could have been written by anyone (or a computer); there is no individual vision; the text lacks an abundance of revealing details; it is full of generalizations with no documentation. The draft is not satisfying to read; the reader learns nothing the reader did not know.

────────────── ■ *Writing Exercise* ■ ──────────────

Take a piece of your writing and write more in the margin wherever you would like more information. Do the same with published and unpublished writing you like and do not like to see how a well developed piece of writing serves you and an undeveloped one does not.

■ TECHNIQUES OF DEVELOPMENT ■

An effective piece of writing has depth that goes below the surface. The effective writer develops a richness of texture that attracts and holds the reader. *"We suffered a flood"* becomes *"I expected a huge, sudden tidal wave, wind, rain; but what we experienced was worse: the quiet arrival of the river that usually was a mile away—first a trickle; then streams and puddles; then the quiet, continuous rising of water as far as I could see; my horizon rising and trees, telephone poles, houses, barns all sinking out of sight—quietly, certainly."*

There are three basic problems with many early drafts which can be solved by development. One is too little information.

Another is lack of authoritative documentation. And the third is lack of clarity. Solving these problems, however, does more than satisfy the reader. It educates the writer.

Develop with Information

Reveal with Specifics

The writers we read, enjoy, trust—the writers who inform and stimulate us, write with information: revealing, specific, accurate, interesting information. Non-writers think we write with words, brightly colored balloons, empty of information; but words have no value unless they are loaded with information. Words are nothing in themselves. They are devices to carry information, the way a check is a device to deposit, withdraw or transfer money in the bank. No money, the check bounces; no information, language bounces.

Write with Abundance

Meaning, form, order, voice lie within the material. Developing a draft is an intellectual act of enormous importance. We don't think of writing with telegraphic summations—environment should be saved—but with a rich plenitude of information about a specific threat to the wetlands and the details of complex problems that might be caused by a specific solution.

Develop with Authority

Convince with Authority

It is specifics that give the writer authority with the reader. The reader is impressed by an abundance of information that convinces that the writer knows what he or she is talking about.

Persuade with Evidence

Most writing is argument. We want to persuade the reader that our vision of the world is true. The reader should share our opinion, follow our leadership, vote for our proposal, buy our product. To persuade the reader, the writer must deliver evidence that the reader will believe.

Personal Documentation. Some of the evidence may be personal. In writing about date rape, a student described what happened to her roommate, a friend from home, a sister, herself. The reader needs to know where the documentation is coming from. **Attribution**—telling the reader the source of information—is vital in persuading the reader, who should be suspicious of information that is not connected to an authoritative source.

Objective Documentation. Most evidence comes from impersonal sources. The reader of the article on date rape demands—fairly or not—to know whether this is a problem of the individual writer or whether it is a widespread problem.

The writer should build her case with an abundance of information that comes from campus and local police reports, medical records, campus housing administrators, sociological students, case histories of victims. Again, attribution is vital. The reader needs to know from statements within the text, or from footnotes, where the writer has gotten the documentation.

Develop with Clarity

Description

Description is the mother of all writing. With words we describe our worlds, physical and intellectual. We describe observations, ideas, theories, people, events, concepts, experiments, emotions, processes—the range of human experience.

Two essential elements of effective description often forgotten by inexperienced writers include:

Dominant Impression

The description has a focus, everything in the description supports a single impression or meaning.

> Everything in the operating room was sharp—piercing light glinted off stainless steel equipment with clean edges; the sounds of machines echoed brightly off the tile walls; the eyes of masked doctors, nurses and technicians probed the patient; a tray held a battalion of knives, each sharpened, pointed, waiting.

Natural Order

The reader should receive the description in a natural order, a story built on chronology; time passing; a description of an argument moving from the weakest to the strongest points; a place seen in the order the narrator would see it.

> The first thing she felt as they wheeled her into the operating room was the sharp piercing light that glinted off stainless steel equipment; then she became aware of the clean, precise edges of the equipment, the clatter of tools and the pulsing beat of the machines that would breathe for her; the piercing eyes of masked doctors, nurses and technicians; and before the mask came over her face, a glimpse of a tray with a battalion of knives, each sharpened, pointed, waiting.

Put Meaning in Context

Many writers deliver information—facts, quotations, case histories, reports, quotations—that are interesting but cause the reader to ask "So what?" Pure information is not enough; that information has to be put in context. The anecdote or personal experience must be put in a larger context for the reader to understand its full meaning.

───────────── ■ *Writing Exercise* ■ ─────────────

Since it is easier to see what someone else's writing needs than to see the needs of your own, take a piece of writing by another writer on a subject you know well and mark in the margin how each paragraph or section that needs it, could be developed.

■ RE-WRITE WITHIN THE DRAFT ■

When writers revise their drafts, they are tempted to look beyond the page to identify problems and solutions. They try to remember what their teachers or editors have said about this form of writing. They study their notes and look into writing texts (perhaps even mine), consult with friends, family, classmates; but once the writer is within the draft, the most important place to look is within the draft itself. If you read the draft as a stranger—reading what is and what is *not* on the page—the draft will often tell you what it needs.

Here are some examples of student writers engaging in a dialogue with their drafts:

> High school was, to me, was even more boring than home. I'm only in college because of the job I had. That really changed me.

> *How was school and home boring? What job? What change?*

> After school, was when my education started, Okay, I suppose in the morning too, when I had to prepare for work. At night I really started doing my homework, not for school, but my job. But in the afternoon, I wasn't a kid, I was a working girl.

> *Straighten out the chronology. How about going through a day?*

> Paper route, 126 AM papers, used fifth hand Pontiac I bought from Uncle Jim who was in the Coast Guard, married to my mother's younger sister, she was band director, went to that before school, played tuba cause I was the small guy, I suppose, tuba with feet they said, big noise, played jazz tuba, too, don't laugh, and then school where I had math first period, teacher had a rug, no rugs, he was redhead-blond-prematurely gray, it depended, his wife was the shop teacher, had six fingers—on both hands—no, two on one, four on the other, and . . .

> *Whoa. Slow down. What do these specifics mean? Where are you taking me?*

Development during revision is often a matter of working through the draft, paragraph-by-paragraph, line-by-line, sanding, fitting, rebuilding, shaping, caulking, adding, cutting, according to the demands of the line.

Boring? No. Because as you develop the draft, it grows and changes, teaching you what you don't expect about your subject.

▪ EMPHASIZE THE SIGNIFICANT ▪

In developing a text it is important to make sure you are giving the information you are adding the proper emphasis to make your meaning clear. In accumulating an abundance of information there is a danger that the draft will be piled up with so many lists and heaps of information that the reader will not be able to see its significance.

Emphasis, of course, is provided by the dramatic nature of certain pieces of information, by the vigor of the writing but, most of all, by the placement of the information. Where you place the most important information in the sentence, the paragraph, the section, the article, the book is important.

This can be seen most easily in the paragraph:

2nd Point of Emphasis
[important information that will attract reader]

3rd Point of Emphasis
[attribution and other less interesting information that needs to be included]

1st Point of Emphasis
[most important information. Will stay in reader's mind and make reader read on.]

Often the most important information is buried in a paragraph and the reader zooms right over it. When you have information that must be emphasized you should realize the reader remembers best what is at the end of the paragraph, next best what is at the beginning, and least what is in the middle of the paragraph.

This 2-3-1 principle works in key sentences and in larger blocks of writing, but it should not be followed all the time, as this will cause each sentence, paragraph and piece to sound the same: Ta-Boom, Ta-Boom, Ta-Boom.

In revision, I consider the 2-3-1 question when I read significant pieces of writing that must be clear to the reader, but which are totally confusing. Often the important material is buried in the middle, and moving it out to the edges will clarify the writing.

─────────────── ■ *Writing Exercise* ■ ───────────────

Take a paragraph of yours or someone else's and move the points of emphasis around to see how that affects the draft.

Remember, there may be good reasons to put the key information in the middle of the paragraph, or at the beginning. Do not follow this advice slavishly, but consider it if test readers are confused.

■ PACE AND PROPORTION ■

Two important elements in writing that are often ignored in writing texts are pace and proportion.

Pace is the speed at which the writer causes the reader to move through the text. One way to speed up the pace is to use short sentences and sentence fragments, what one person calls "the English minor sentence": *"Now is the time to vote. Not in the next election. Not tonight on the way home from work. Not during lunch. Right now. On your way to work. This day. This hour. Right now."* The pace can be slowed down by longer words and longer sentences with clotting clauses: *"It is understandable that citizens with multiple responsibilities procrastinate and wait to vote at a later time period that never comes. There is always a meeting, an overdue assignment, a deadline project, a crisis—real or imagined—that delays electorate until the vote that affects their lives, and the lives of their families, cannot be counted because the polls have closed."*

Proportion is the relationship of the parts to each other. Description may need to be balanced by dialogue, facts with people, theories with evidence.

As you develop your draft, you will confront two key questions: How fast is fast enough? How much is enough?

The issues of pace have to be solved together, for when you speed up or slow down a draft, you alter the length of each section and when you decrease or increase the length of a section, you speed up or slow down the reader's pace.

Proportion is a matter of length but the decision as to what is enough exposition or description is affected by the length of other parts. A detailed description of a manufacturing process might allow a detailed analysis; a quick anecdote about the designer of the process might limit a quotation to a sentence. How much ketchup depends, in part, on the size of the burger.

The re-writer wants to give the reader all the information the reader needs and no more. If there is too little, the reader will stop reading; and if there is too much, the reader will stop reading.

Pace is not only influenced by length but by the way the piece is written. Long sentences and paragraphs slow readers down; short ones speed them up. Dialogue increases velocity; description slows it down. Verbs and nouns accelerate; adverbs and adjectives apply the brakes.

There is no right pace. It depends on many factors, such as how familiar the readers are with the subject, how specialized their knowledge, how much the writer wants to entertain or inform, to appeal to the brain or to the heart.

------------------------------ ■ *Writing Exercise* ■ ------------------------------

Take a draft or your own and play with pace and proportion. This is always play, quick experiments to see what works and what does not work—for the reader.

■ REVISING ACADEMIC WRITING ■

In developing academic prose the writer must give the reader persuasive evidence that is documented. Academic prose appeals to the intellect more than to the emotions but that does not mean it is dull, distant, generalized, boring. Academic writing can achieve the highest goals of fine prose.

The territory of academic writing is the battlefield of ideas. Academics are doing nothing less than battling for control of their readers' minds. In academic writing you see the sword play between opposing beliefs, theories, concepts. Idea attacks idea as critical minds, disciplined by print, parry and thrust.

To do this, language must be precise and accurate, and to hold the interest of the reader it must be entertaining, lively, sparked by wit. Grace is as important in academic writing as in any other form of composition.

Academic readers, of course, demand substantial, accurate information, attributed to reliable sources. They want to know what supports the writer's opinion and where it comes from.

Some inexperienced academic writers think it is possible to build academic writing out of lofty generalizations founded on invisible assumptions. It is not. Academic writing must be specific. It is not enough to list information, however. The specifics must be put in context.

Academic writing is thought laid bare. The writer has to analyze the information, then build a meaning from it. Academic writing is critical thinking written down. It establishes ideas, options, theories, propositions, arguments, and attempts to persuade by a combination of evidence and thought based on that evidence. The writer has to make sure the draft has specifics but also well founded and developed generalizations that give the specific meaning.

■ NOTES ON WRITING THE PERSONAL ESSAY ■

The personal essay, if it is successful, looks easy, and it should; but the craft that produces easy writing has its own demands. Recently I was asked to speak to a group of professional writers who wanted to publish personal essays and I wrote the following notes:

- *Respect yourself, your personal vision.* Cultivate an aware self-centeredness of the world around you and the world within you. You have your stories to tell—and a responsibility to tell them. Value what you catch out of the corner of your eye, hear from the next booth or from your own mouth, what you say when you talk to yourself, think when you are not thinking. Pay close attention to the obvious.

- *The more personal you are, the more universal you will be.* It is the mission of the writer to articulate the inarticulate thoughts and feelings of the reader.

- *The personal must not be private.* The personal must be in a larger context. The essay must have a significance beyond your life, an opinion on the meaning of your experience for you—and your reader.

- *Write with revealing, specific details.* You have seen more than you remember until you write. Use these recovered specifics that resonate for the reader. Specifics make the writing lively, grant you authority and, unexpectedly, provide universality.

- *Start as near the end as possible.* Don't tell the reader what you are going to do; do it. Weave in background information at the moment the reader needs it.

- *Take your reader on a voyage of discovery.* Write a narrative that reveals what you didn't know you knew. An essay works when I am surprised half why through with an unexpected meaning I share with the reader.

- *Start with a line, not a subject.* Start with a fragment of language or a haunting image that contains a tension, conflict, contradiction, irony, problem: write to discover what you have to say.

- *Write with velocity.* Write fast to outrun the censor; force significant accidents of meaning and expression; and create instructive failures.

- *Write out loud.* Writing is read and believed because of voice, more than any other element. Write out loud, hearing your draft as it heads toward the page, tuning your natural voice to your content and to your reader so that each text has its own consistent voice that will be heard by the reader.

- *Say one thing.* The personal essay must focus on and develop one thing. Texture comes from the developing material.
- *Less is, indeed, more.* Eight hundred words is always better than twelve hundred. Cut anything that does not move the draft toward meaning. This does not mean writing only in simplistic declarative sentences, but every pause, every detail, every change of tone must move the reader forward.
- *Don't compress; select.* Brevity is not produced by a garbage compactor. Select the anecdote, idea, scene, opinion that is central. Develop and document it fully.
- *Don't tell; reveal.* Do not insult readers by telling them what they should feel or think. Give them the specific material by which they will create their own text out of their own needs and their own autobiography—and find in it their own meaning.
- *Acceptance is as irrational as rejection.* You are in a personal, irrational business. Do your work. You will learn little from acceptance, less from rejection.
- *Cultivate a good editor.* The good editor helps you write more like yourself. Tell the editor what you need; thank the editor if you get it.
- *Be careful who you allow to read your drafts.* Most readers—and editors—have expectations of what you *should* say and how you *should* say it: who you *should* be. Seek readers who help you write like yourself—and make you want to get back to your writing desk when you leave them.

■ *Writing Exercise* ■

Adapt the notes above that I made for professional writers to writing for specific courses. For example, writing chemistry lab reports, history book reviews, nursing case histories, papers of literary analysis, corporate memos, sales letters, term papers, whatever form of writing you have to produce in school or on the job.

A PROFESSIONAL
CASE HISTORY

You can see if I followed my own counsel as I walk you through a personal essay. This essay / column began when Christopher Scanlan, a close friend and a writer in the Knight-Ridder Washington Bureau, wrote a powerful story about the fact that handguns kill five thousand children under the age of nineteen and another thirty to fifty thousand children are wounded each year. Reading his story, I found myself traveling back to the day my first handgun was placed in my hand.

I start with a line—a fragment of language that itches, has tension: a contradiction, a problem—or an image. This time it was the weight of that forty-five that still seemed so natural in my hands. That naturalness and the devastating anecdotes and statistics in the article created a tension in my mind I had to explore in writing.

As I said near the beginning of this chapter, you have to respect your own vision. I am well aware of the fact that there are people who know more about guns than I do, who have used guns more than I have, who have suffered much more combat than I did. But I still have my individual story: the story of a single human being—and each human being's story is worth telling—and hearing. My doubts are similar to the beginning writer's doubts—what do I have to say? Why should anyone listen to me? And I have to overcome them each day.

I write my column Monday morning a week ahead and write fast to get ahead of the censor, to discover what I have to say that I don't know I have to say, to cause the accidents of insight and language that instruct me.

I write the column—1278 words—in about forty-five minutes. Sometimes it is less; sometimes more. Never an hour. I read and edit it twice, submit it to Minnie Mae, my wife and first reader, then send it off to Evelynne Kramer, a superb editor who helps me write as Don Murray. Many editors try to make you write like someone else but Evelynne wants me to write like Don Murray.

She called with two concerns: first, a general one that I am writing about World War II too much, using excessive high drama to make points. She fears readers of my over-sixty column—sixty to seventy percent of whom are under sixty—will say, "Oh, Don's going back to the war." Also she thought I could tighten the piece up, wondered whether the dialogue works and suggested I'm repetitive about the "power" of the gun.

Writing can always be improved by cutting. To emphasize the significant, you often have to cut good material, as you will see in this case history, which pulls the reader away from focussing on what is most important.

When my editor mentioned cutting, I marked the scene at the filling station reprinted below as a possible cut.

I also did a word search on the computer and was shocked to discover I used "power" seventeen times, "powerful" twice!

Then I went to work revising, cutting line-by-line and eliminating. I cut the filling station scene I had marked. It also contained the dialogue she questioned. This became a matter of proportion. I didn't need this scene. It was interesting autobiography but wasn't necessary to my meaning and slowed the pace of the essay down. Here's the scene I cut:

> When I was in high school, hanging out at the Amoco station, leaning up against the wall, bragging about what I had not yet done, bored with school, bored with life, an outsider wanting to be an insider, some kids drove up, tough kids, kids I feared, despised and envied.
>
> "Hey, Murray, get in."
> "Nawh."
> "We're gonna ride around."
> "So?"
> "So come along."
> "Nawh."
> "Why not?"
> "I dunno."

And I don't know why I didn't go with them. I didn't know they had a gun but I certainly wanted their acceptance and just being with them would tell others I was tough. I was saved, I suppose, by the incredible *power* of teen-age inertia. They held up a gas station, killed the guy who owned it, allegedly with a forty-five like the one the sergeant lay in my hand.

I get a great deal of satisfaction and no sense of failure in making such a cut. It was a good yarn, reasonably well written. There is no way I could tell it was not necessary without writing it and reading it in relation to everything else in the essay.

I also delight in the craft of my line-by-line editing. As I cut, the essay becomes smoother, easier to read, and more powerful. I release it from rhetoric. Here's an example of line-by-line editing.

I was eighteen years old and I had ~~power in my hand. Raw power. Primitive power. Magical power, the power of the comics, the power of legends,~~ the power King Arthur's Knights of the Round Table felt in their swords, ~~the power~~ Robin Hood's Sherwood Forest archers felt in their long bows, ~~the power~~ Buck Rogers felt in his space zapper, ~~that~~ Tom Mix felt in his six-gun. ~~I had been initiated into the manly blood rite of guns. And~~ this heavy handgun felt amazingly light in my hand.

Sometimes you have to write badly to write well and I certainly wrote badly here. In the published column, this passage reads:

I was eighteen years old and in my hand I held the power of the sword, the long bow, the six-gun. And it was light in my hand.

Much, much better.

That rough cut brought the draft down to 919 words from 1278. I felt it was still a bit long. I changed some things in cleaning up the editing and read it again, cutting it down to 882 words. Then another read. Up two words to 884. I sent it in. 1294 Words − 884 = 410 cut from 1294, 31 percent.

Evelynne Kramer liked it and here is my personal essay as it ran.

It is fifty years ago this spring that my fingers first curled around a handgun.

The army sergeant had me hold out my right hand, palm up; then he ceremonially lay a forty-five-caliber automatic pistol on my hand and smiled as my fingers naturally curled around it.

"Naturally" is the key word. I give you the dominant impression of the gun and then reveal my hand's reaction to this weapon. I simply wanted to recreate the image, to get out of the way of the writing. And then I go on to establish my authority to write on this subject by providing the reader with specific evidence that I know what I am talking about.

In my army service I was trained—marksman or better—on three kinds of rifles, two carbines, three submachine guns, light and heavy machine guns, an old fashioned "six-gun" revolver, even a rifled shotgun, but nothing felt as good, as natural, as that forty-five.

Notice how I instinctively followed the 2-3-1 paragraph design to emphasize the forty-five. I also want to increase the context and place it within the fantasy world of the young boy—girl?

I was eighteen years old and in my hand I held the power of the sword, the long bow, the six-gun. And it was light in my hand.

The line that started me writing the essay might have been, "light in my hand." It contains a tension between the 'heavy' possibility of the weapon and lightness or ease with which I wield it.

With a little training I hit the target in or near the heart every time. We did not shoot to injure, we shot to kill.

"Don't draw your gun unless you intend to use it. Don't use it unless you shoot to kill."

Answer reader's questions when they are asked. Reader asks: "Did you?"

One of my assignments was as an M. P., a military policeman. And more important than the arm band, all the paraphernalia we strapped on, the legal nightstick, the illegal slapstick, was my forty-five.

Within the narrative a writer can work in exposition—how it felt— implication of how it feels to kids on the street.

On patrol, my fingers kept straying to my holster, to the butt of my gun; I felt its pressure against my hip like a caress. I was not a kid. I was man, somebody to reckon with, someone in control, someone mean and tough and hard.

But would I have the courage to use it?

Write long then short. Short for emphasis.

It was easy.

A fellow paratrooper, drunker than a coot, broke a bottle on a bartender's head in a Tennessee road house and ran out, across a plowed field.

> *I develop the anecdote I use with abundant detail even if the essay is short. We write short by selecting carefully, then developing what we leave in. Had I left in the filling station anecdote, I would have less room to develop this anecdote.*

I took after him. It was easy to pull out my forty-five, easy to aim and shoot at him running across the uneven ground. Stupid. I missed.

> *I weave in my basic training instructions. This is pure narrative. Puts the reader within the "I." And note how I follow the natural order I discussed earlier in the chapter. In this case it is a chronological order, a chase, something the reader is familiar with.*

"Don't draw your gun unless you intend to use it. Don't use it unless you shoot to kill."

Stop. Kneel. See him against the sky. Aim. Fire. Another miss. But close. He stopped.

He had murdered the bartender, but I didn't know that when I fired. I swaggered in with my terrified drunk and felt great. I was a man for sure.

> **"Within my head at nineteen"**—*reinforces what young males who carry guns to school feel. Note the visual quality of the verb "swaggered."*

I used most of the other weapons in combat and along the way got hold of an army forty-five illegally—even had a shoulder holster made for me.

I echo back to the third paragraph, hoping the seed I planted in the reader's mind has survived.

On the troop ship home, my forty-five and I earned good money as a bodyguard to a kid from New York who ran a crap game. Don't laugh. I heard of at least one combat "hero" tossed overboard on our trip home.

Implies not just combat but a social problem. This writing of the draft was instinctive, but now, during the editing, what I have done on instinct has to be considered rationally:

What works?

What needs work?

What doesn't work?

I never used a forty-five to shoot at another human being after that night in Tennessee, but I loved to carry my secret weapon.

Now the lecture. Have I earned the right to do it? A more important question: Will I be more effective if I do not lecture?

Let's witness to true feelings. With my handgun I had nothing less than the power to kill. Mess with me and I can blow you away.

Turn up the heat on the handgun-owning reader.

And—surprise—you don't know I'm carrying. A handgun is usually a hidden weapon, in the belt, in the pocket, in the purse, in the bedside drawer, in the glove compartment.

Turn it up some more.

And it is loaded. No sense in having a gun locked in one room, the bullets in another. What if an intruder . . .

Now connect the handgun-owning reader with the kid on the street.

I am not shocked or surprised kids are carrying—and that many of their parents have their own handguns in the bedside drawer or glove compartment. I was intimate with the seduction of the gun and I have to live with my love of that power that lay so easily—so naturally—in my hand.

Weave back to "naturally" and focus on me so reader is not forced into a corner, into a defensive position. We are human beings and that is the problem.

I turned my forty-five in when I was discharged, to the surprise of the sergeant, who probably never reported it and took it home. I don't know why I left it behind and at times I regretted it. Not now, but in the first years after the war when I was young.

Explicit statement of position. Does it weaken piece?

I am for federal handgun control, for making it illegal for anyone except the police or the military to own or carry a handgun, illegal to manufacture handguns and sell them to anyone but the police or the armed services.

Our nation suffers nothing less than a civil war in our streets, in our schools, in our homes: a war fought with handguns. We should buy back, search out and confiscate every single privately owned handgun.

If you own a handgun you share—privately, secretly—my seduction by the gun.

Nothing less than mind invasion.

Everyone has his or her own fantasy: the late night intruder invading my home, the gang member daring to walk my turf, the hold-up man in my store, the man who took "my" woman, the driver cutting me off on the lonely highway, the rapist. No matter the reason, justified by the court or unjustified, the target of every fantasy is the same: another human being.

"Don't draw your gun unless you intend to use it. Don't use it unless you shoot to kill."

***

STUDENT CASE HISTORY
Chris Dufort

■═══■

Chris Dufort's piece grew out of a class exercise in which he wrote an authority list, recording things on which he was an authority. Then a small writing group helped him see what topics on his list might interest readers. He describes his process in writing this paper:

> The paper started with a quick free write in which we had to pick something off the top of our heads and write. The paragraph I had was terrible but everyone liked the idea of a rooster being vicious.

The first paragraph he writes shows how an idea comes in the writing. He writes:

> The animal rescue league was an awful place. There were so many things I hated about it. I hated the people that came in and the things they did. I hated the cleaning. The rooster. I doubt there was much I liked . . .

To someone who does not write, that paragraph does not seem promising. It is written in generalities. But look at what happens when the writer pursues a specific *"the rooster."*

> The rooster stalked slowly. When I had heard the warnings: Wear boots, bring a broom, I had only scoffed. It's only a rooster I said. They just smiled and sent me out knowingly. I went out to the barnyard naive of what lay waiting for me. The rooster.

Now we are intrigued. The writer has a subject and Dufort knows it. He writes:

> The paragraph I had was terrible but everyone liked the idea of a rooster being vicious. Next we picked a focus and were supposed to write three leads. I wrote one. I knew it was good and would need no more. I used it that week for my paper. It contained my lead unchanged. I liked the start of it but nothing else. It changes in the middle and goes nowhere. To revise it I followed one suggestion to read it four times. I began to realize other things about the piece and my feelings and went from there.

His first draft demonstrates the accuracy of his own evaluation. But he should not forget that the writing that did not work was essential: it led him to the rooster and then to the real meaning of the essay.

The rooster stalked slowly. When I had heard the warnings: "wear the boots," "you'll need room to defend yourself," I scoffed. It's just a rooster I said. They just smiled and sent me out knowingly. I went out to feed the barn animals naive of what lay there waiting for me. The rooster.

Entering the yard I looked around. The League had taken in a lot of farm animals over the past few years. They had a few goats, some geese, several hens, a duck, even a Canada goose. And the rooster. I quickly looked around for it again. It wasn't anywhere to be found. For the time being I was safe.

Looking around I remembered what brought me to the League in the first place. Laziness. I was fresh from being kicked out of college with a .61 grade point average and being forced into a job by my parents. I just wanted to continue the lifestyle that had led to my lackluster school performance. I wanted to do nothing. I had to go back to school in seven months. Assuming they let me back in. I had to prepare.

I started looking through the paper for jobs that appealed to me. Cooking, cleaning, stocking shelves, it was all there, but that stuff all sounded like work. Then I came across an ad from the Animal Rescue League. What could they possibly have for work? Feeding and petting animals? That sounded good to me. I called to see if they were still hiring and then went to get the application. Soon I had an interview with the director of the League. She explained to me that they were part of the Humane Society and took in stray and unwanted animals and put them up for adoption. With a little bit of lying and a lot of smiling I got the job. A job taking care of animals! And it was only part-time. God had obviously smiled on me.

The first day all I did was clean shit.

The second day all I did was clean shit.

This was definitely not the job I had been looking for. All I did was clean out cages. I cleaned out cat cages, I cleaned out dog cages, and when I was done cleaning out cages it was time to go home. Where was the fun? Where were the animals? This just sucked. It was getting ridiculous.

It all ended in a flurry of feathers and clucking. I felt the rooster's spurs dig into the back of my legs. It hurt. The stupid bird hurt me. And it did it from behind. And it was now between me and the door. I looked at it and it cocked its head, staring at me. I didn't know what to do, but boy was I pissed. I threw food at it. It cocked its head at me a little more. The dumb beast didn't move. I stomped my feet and yelled at it. It cocked its head at me a little more. This was just stupid. I had to get out the door but I couldn't get by the rooster without getting hurt again. Out of the corner of my eye I saw a rock. I took a step to my right closer to it. The rooster turned to face me. All of sudden I was nudged from behind. As I turned to face my new assailant the rooster moved forward. I turned to face it and it stopped. I threw some more food at it just to be sure. As if things weren't bad enough, I now had a goat nudging me from behind to contend with. At least these things were really dumb. As the goat moved next to me, I took another step to my right. The rooster, not liking all this confusion, charged at me. I pushed the goat in its way and grabbed the rock. I threw it. I missed. The rooster charged again. This time I kicked it out of the way and broke for the door as fast as I could and locked it behind me. Round one to the rooster.

The following days were more of the same. More cleaning the cages and more feeding the animals. More early mornings and more rude people. They even had me mowing the lawn now. The only thing that kept me there was the hassle of finding another job. I sometimes wondered what my friends at school were doing. I began to think about going back.

I began to think about going back into the barnyard again also. So far whenever I had to feed the barn animals I just leaned over the fence and threw food in, but I knew I couldn't get away with my laziness forever. Soon it would catch up with me and someone would notice. I began to appreciate cleaning up after the other animals.

The next day it all caught up with me. I was taken aside and told that I had to feed the barn animals properly. This time before I went in, I put on the knee high boots to protect my lower legs and armed myself with a broom. I entered and moved along with my back to the fence. As soon as the rooster heard the gate open it came out of the barn. I moved towards the food trays for the animals and the rooster moved towards me. But it didn't charge. It stayed between me and the hens, but it kept its distance. It seemed to be afraid of the broom. Someone must have had a little fun at the rooster's expense one time. It had learned its lesson.

At school now I sometimes think about the rooster. Sometimes I think about the League. I still hate them both. I guess I always will. That can't be lied about. I think the only thing I regret now is that I didn't get there earlier than I did so that I could have seen the fight that made the stupid rooster afraid of the broom.

My writing follows the same pattern—a good start that peters out. Not to worry. There is always another day. I edit the early part and then when I get up to speed I find myself pushing on as Dufort did as he developed his essay. Here is his fourth draft. It may not be his final draft, but it is more fully developed and better shaped, and he increasingly puts his personal experience in a larger context. He reveals what the experience taught him. Writing can always be improved. But, as you will see, he has come a long way from that first lead paragraph.

The rooster stalked slowly. I had heard all the warnings about wearing boots to protect my legs, about bringing a broom to fight with, but I only scoffed. It's just a rooster I said. They smiled knowingly and sent me off to my doom. I went out to feed the barn animals, naive of what laid there waiting for me. The rooster.

Entering the yard I looked around. The League had taken in a lot of farm animals over the past few years. They had a few goats, some geese, several hens, a duck, even a Canada goose. And the rooster. I quickly looked around for it again. It wasn't anywhere to be found. For the time being I was safe.

Looking around I remembered what brought me to the League in the first place. Nothing but pure laziness. I was fresh from being kicked out of college with a .61 grade point average and being forced into a job by my parents. I enjoyed the life that led to my lackluster school performance. I enjoyed doing nothing. I had to go back to school in seven months. Assuming they let me back in. I had to prepare.

I started looking through the paper for jobs that appealed to me. Cooking, cleaning, stocking shelves: it was all there, but that stuff all sounded like work. Then I came across an ad from the Animal Rescue League. What could they possibly have for work? Feeding and petting animals? That sounded good to me. I called to see if they were still hiring, then went to get the application. Soon I had an interview with the director of the League. She explained to me that they were part of the Humane Society and took in stray and unwanted animals and put them up for adoption. With a little bit of lying and a lot of smiling I got the job. A job taking care of animals! And it was only part-time! God had obviously smiled on me.

The first day all I did was clean shit.

The second day all I did was clean shit.

This was definitely not the job I had been looking for. All I did was clean out cages. I cleaned cat cages, I cleaned dog cages, and when I was done cleaning cages it was time to go home. Where was the fun? Where were the animals? This just sucked. It was getting ridiculous.

Suddenly in a flurry of clucking and flying feathers my reminiscing ended. The rooster was upon me. I felt its spurs dig into the back of my legs. It hurt. The stupid bird hurt me. And it did it from behind. And now it was between me and the door. I looked at it and it cocked its head, staring at me. I didn't know what to do, but boy was I pissed. I threw food at it. It cocked its head at me a little more. The dumb beast didn't move. I stomped my foot and yelled at it. It cocked its head at me a little more. This was stupid. I had to get out the gate but couldn't get by the rooster without getting hurt again. I swallowed and ran at it. It spurred my leg again, but this time I kicked at it and got out the gate. I shut it behind me and leaned on it. I still had the bucket of food for the animals. I leaned over the gate and dumped it in. I vowed to never enter the barnyard again.

Working at the League was a new experience for me. Probably because that was exactly what it was. Work. I had never experienced all the trials and tribulations that go along with work. I also never had responsibility. I never had the life of something depending on me.

One time I was asked to hold a dog while it was put to sleep. The shot they give it burns a little and they need someone to hold it down. They also like to have someone to comfort the dog as it dies. The dog didn't seem to mind the shot at all. It just lay on the table with its head on its front paws. I looked into its face confused. I had never seen anything die before. I felt weird. Its eyes looked calm. Knowing. Even sad. Slowly the spark in its eyes went out. It was as if the life inside that fanned it was gone and now it went out. I looked around at everyone in the room. They were petting the dog. Telling him how good he was. How nice he was. Saying his name. They took it so calmly. I felt everything rush in on me. I wanted to scream. Couldn't they see this? It was dying. I couldn't believe how careless they were. How unfeeling. I looked at the director in time to see a tear run down her cheek. I realized then that they did feel it. Perhaps even more deeply than I did. I wanted to cry too. Later as I loaded the cold body into the furnace to be cremated, I did.

One time I got a raise. I was told that it was because it was policy to review everyone and give them a raise after three months. They told me that I got the raise because they had to. Not because I deserved it.

I began to discover that the animals here did depend on me. That their lives, the lives no-one else wanted, depended on me and the work I did. Once I forgot to lock the dog cages outside. Later in the day I heard yelling and went to investigate. Three dogs got out and went into the barnyard. They were trying to catch the geese and hens. One had bitten the duck. The dogs were finally rounded up and put away, the duck brought in and examined. The only reason I wasn't fired was because I had two weeks left.

Eventually it was time to go back in the barnyard. I went in cautiously, armed with a broom, protected by hip boots. The rooster never came near me. It seemed to be afraid. Probably of the broom. I smiled, imagining the battle in which the rooster learned fear.

The last day I worked there I got to go on a trip. The League was often left with raccoons that people found and they are taken care of until they can be set free in the woods. We got in four baby raccoons once. They were given to me to be taken care of. I fed them by hand and then, when they were bigger, I cleaned out their pen and fed them canned cat food. On the last day I worked they were to be set free and I got to go. We carried them in small cages a mile into the woods. We opened the cages and backed off. For a long time they didn't come out. Then slowly, one by one, they came out. I swelled with pride as they played and began to leave. I did this. I was the reason they were here to be set free. I saddened, realizing I wouldn't see them again. They went off. One climbed a tree and I felt proud again. I wonder now if this was how my parents felt when I went to school. I can only imagine how they felt when I was sent back to them.

At school this year I almost made the Dean's list. I was .03 short. I was told at my new job that I will be getting a raise faster than anyone else. When I tell my parents this I know how they feel. I can hear the pride in their voices when they ask me to tell people of my accomplishments. I want to tell them that I can relate. Sometimes I want to tell them that I was taught to feel the same thing by four raccoons.

Note how each stage of the revision process has made you re-think. Writing is a process of learning and re-learning. In fact, for many of us, it is the most effective way of learning. As we provide the reader with documentation, we see the subject with increasing clarity and may have to change what we think, feel, and therefore, communicate. The process of self-education continues as we polish and clarify what we have to say in the final stages of the revision process.

CHAPTER 9
RE-WRITE WITH VOICE

Now the fun begins.

Now I can play the music of language that will wrap around the words and give them that extra aura of meaning that is the mark of effective writing. It is the music of the language that draws the writer to the writing desk and informs the writer of the meanings and feelings that lie within the subject; it is the music of language that attracts and holds the reader and causes the reader to trust and believe the writer; it is the music of language that provides emphasis and clarity; it is the music of language that makes the writer and the reader *hear* the printed word.

Now, at this crucial moment in the revision process, I have discovered meaning; I have selected convincing evidence; I have read my readers and discovered their needs; I have chosen an appropriate form and constructed an order within it; I am ready to sing.

We introduced and defined voice on pages 8-11 of the first chapter and now we have the opportunity to focus on what is the most important quality of writing that is read.

▪ DIAGNOSIS: NO VOICE ▪

Lack of voice is the most common reason we stop reading, but we do not name it. No reader says, "I abhor the lack of voice in the pages" and puts down the article or book. They simply stop reading. Their minds float off the page. They realize they are not reading and turn to something else. They have heard no music even if they do not realize there should be music. In the same way, readers who read on do not say, "I am being carried forward on the music of the writer's voice." They just read on, hearing and not knowing they are hearing a voice rising from the page.

As writers, however, we must become aware of the voice of what we write or of the lack of a voice. The signs of a text without voice include:

- *No individual human being behind the page.* The page could have been written by anyone. The author is anonymous. The text was not created by a living human animal but by a machine.

- *No intellectual challenge.* The page does not make intimate combat with the reader's mind. It does not stimulate, challenge, inform, surprise the reader, inspiring the mind-to-mind combat that marks good writing.

- *No emotional challenge.* The page does not engage the reader's emotions, forcing the reader to feel as well as think.

- *No flow.* The reader is not carried forward by the energy of the voice that connects all the elements of writing into a powerful river of language that makes it hard for the reader to escape the page.

- *No magic.* Good writing gives the words more meaning than they have while lying separate from each other in the dictionary. The magic of writing is that the voice rises from the spaces between the words as much as from the words themselves carrying meaning to the reader.

Writers write with their ears, listening to the music rising from the page or from the computer screen as they write and re-write.

■ HEARING THE WRITER'S VOICE ■

Once we are aware of the importance of voice, we should read good and bad writers to *hear* the music of their texts—and the marvelous diversity of human voices writing in every genre, for every purpose. Here are some examples of voice I admire—and the reasons I admire them.

> *Here is an account of a few years in the life of Quoyle, born in Brooklyn and raised in a shuffle of dreary upstate towns.*
>
> *Hive-spangled, gut roaring with gas and cramp, he survived childhood; at the state university, hand clapped over his chin, he camouflaged torment with smiles and silence. Stumbled through his twenties and into his thirties learning to separate his feelings from his life, counting on nothing. He ate prodigiously, liked a ham knuckle, buttered spuds.*
>
> *His jobs: distributor of vending machine candy, all-night clerk in a convenience store, a third-rate newspaper man. At thirty-six, bereft, brimming with grief and thwarted love, Quoyle steered away to Newfoundland, the rock that had generated his ancestors, a place he had never been nor thought to go.*
>
> E. Annie Proulx
The Shipping News

This novel had been recommended to me and when I saw it in a bookstore this week I picked it up and read these first three paragraphs. I closed the book and bought it. This one I must read. What caught me was the writer's voice. It was individual, powerful, eccentric (another word for individual?) and entertaining. I want to hear this voice spin me a story.

***

> *In many U.S. school systems, there is a curriculum director whose job it is to puzzle out what a curriculum is. The etymology of the word is promising: it comes from the Latin word* currere, *"to run," and is closely related to the word* curricle, *a two-horse chariot used for short races. Presumably, curricles went around in circles just as curricula trends do in this country, the only difference being that curricle drivers knew they were always going over the same ground, and we often don't. The curriculum director, and those who specialize in this murky science in colleges of education, generally tries to keep the chariots moving in the same direction at roughly the same pace.*
>
> *Yet the very order suggested by the word "curriculum"—that fixed track upon which the race occurs—seems antithetical to schooling that*

acknowledges the individual interests and abilities of students. Too often, decisions about what students do at what age are purely arbitrary, and claims by publishers that their work is developmentally sound are only promotional hype. The organization of instruction often shades into regimentation, an interminable forced march through exercises and work sheets. Any misstep, the teacher's manuals imply, might lead to serious problems—like those of the ducklings who didn't learn how to follow their mother at the right time and ended up following the zoo keeper instead.

Thomas Newkirk
More Than Stories—The Range of Children's Writing

Here is a fine example of academic writing in which the author's scholarship is worn lightly and the ideas are presented with clarity, vigor, and grace. This is a seminal study on children's writing, yet there is a place for humor, and you sense the personality of the author from the voice you hear on the page.

***

There is a loneliness that can be rocked. Arms crossed, knees drawn up; holding, holding on, this motion, unlike a ship's, smooths and contains the rocker. It's an inside kind—wrapped tight like skin. Then there is a loneliness that roams. No rocking can hold it down. It is alive, on its own. A dry and spreading thing that makes the sound of one's own feet going seem to come from a far-off place.

Everybody knew what she was called, but nobody anywhere knew her name. Disremembered and unaccounted for, she cannot be lost because no one is looking for her, and even if they were, how can they call her if they don't know her name? Although she has claim, she is not claimed. In the place where long grass opens, the girl who waited to be loved and cry shame erupts to her separate parts, to make it easy for the chewing laughter to swallow her all away.

Toni Morrison
Beloved

These are two paragraphs from the last chapter of Nobel Laureate Morrison's Pulitzer Prize-winning novel. She has the ability to go to the edge of where most of us write and pass

beyond it. Her voice has its own fullness, its own passion, its own richness. It is individual. It is her.

***

To print a document in WordPerfect®, you must have selected a printer and specified the port you were using. This was probably done when your copy of WordPerfect® was installed. If this is the case, you are ready to print.

*You may, however, want to view your current printer selection or change it prior to printing. Or, you may want to use some of the options that are available in the **Print** dialog box before you send your document to the printer.*

WordPerfect® version 6.0

This is a good example of a distant, corporate voice. I have just shifted to WordPerfect® 6.0 and have found that I need other voices to help me.

***

After entering what you feel is the best piece of written work since Tolstoy, you decide that you want to print it. After all, dragging the computer around and showing everyone what your prose looks like on the screen just isn't practical.

To print a document in WordPerfect®—the document you see on the screen (all of it)—do the following:

1. **Make sure that the printer is on and ready to print.**
2. **Press Shift–F7.**

*You then see the **Print** dialog box, a busy place where printing and related activities happen. Don't bother with the details. Just blur your eyes at it and keep reading here.*

Dan Gookin
WordPerfect® 6 for Dummies

This is a fine example of an informal, almost too cute, too friendly writing style, but it was what I needed to get going in

this new, complex software program. Note that these two texts are on the same subject, and so is the one that follows.

***

WordPerfect® offers several ways to print your document. You can print directly from the screen all or part of the document that currently appears, or you can print all or part of a document you previously stored to disk. From the screen, you can print the entire document, a single page, a range of pages, or a marked block of text.

Assumptions

- *A printer is selected in the Select Printer dialog box.*
- *The cursor is at the location where you want to print.*

Exceptions

- *If the page you selected does not appear near the beginning of the document, the printer may pause. WordPerfect® scans the pages for the last format settings for margins, tabs, and so on.*

Trudi Reisner
WordPerfect® 6 Solutions
John Wiley & Sons, Inc., New York, 1993

I needed Dan Gookin's voice to lead me into WordPerfect® 6.0, but now I needed a book that was far more detailed and had a voice that was friendly but authoritative, a voice that was halfway between Gookin and the corporate voice of the official manual. I found it after looking at many books in many bookstores. When I read Trudi Reisner I knew that I had found the office companion I needed.

***

There was a day in my life when I decided to live.

After my childhood, after all that long terrible struggle to simply survive, to escape my stepfather, uncles, speeding Pontiacs, broken glass and rotten floorboards, or that inevitable death by misadventure that claimed so many of my cousins; after watching so many around me, I had not imagined that I would ever need to make such a choice. I had imagined the hunger for life in me was insatiable, endless, unshakeable. I became an escapee—one of the ones others talked about. I became the one who got away, who got glasses from the Lions Club, a job from Lyndon Johnson's War on Poverty, and finally went to college on a scholarship. There I met the people I always read about: girls whose fathers loved them—innocently; boys who drove cars they had not stolen; whole armies of the middle and upper classes I had not truly believed to be real; the children to whom I could not help but compare myself. I matched their innocence, their confidence, their capacity to trust, to love, to be generous against the bitterness, the rage, the pure and terrible hatred that consumed me. Like many others who had gone before me, I began to dream longingly of my own death.

Dorothy Allison
Trash

This is the opening of the introduction to novelist Dorothy Allison's first book of short stories. You can not deny the authority, power, and toughness of this individual voice that writes with such emotion tempered with talent and craft. And notice the specifics that establish her authority.

***

Bao Yanshan's wife was in labour, about to give birth on her bed at home. Big Dog, and son of Baotown's troop leader, ran shouting down to The Lake to find Yanshan. He came sauntering up, hands behind his back, hoe tucked under one arm, thinking what a common occurrence this had become. The seventh belly, no problem, he was thinking—she's just like an old mother hen dropping another egg. To have it come three months early was just that much better; this time of year there was plenty to eat. But whether it was three months or three days, or three hours, it was not worth getting excited about.

Wang Anyi
Baotown

The author's voice comes through translation, and we hear her beginning her book about the remote village to which she was exiled during China's Cultural Revolution. In these few lines her voice takes us into the place far removed from my New Hampshire home, and also takes me into the mind of a person who is my contemporary, yet lives in a different time.

─────────────── ■ *Writing Exercise* ■ ───────────────

Take a few paragraphs of writing you like and write comments on the voice as I have. Then take a few paragraphs of your own and write about them in the third person as I did to help you hear the power and potential of your own voice.

Some people find it difficult to listen to their own voices, to trust their own instincts, to be different from those around them. Parents, teachers and employers have all told them to be somebody else. But if you are going to be a writer, you have to listen to e.e. cummings: *"To be nobody-but-yourself—in a world which is doing its best, night and day, to make you everybody else—means to fight the hardest battle which any human being can fight; and never stop fighting."*

■ HEARING YOUR VOICE ■

At each stage of the re-writing process we have woven in the role of voice, but now, as we edit the final draft, we concentrate on the voice of the draft.

The writer writes with the ear, hearing what is on the page and what is not yet on the page: the tune of meaning is often heard, then played. The writer listens for the writer's own voice, then tunes it to the draft so that the human music of voice supports and advances meaning.

What Do You Mean by Voice?

Voice is the magical heard quality in writing. Voice is what allows the reader's eyes to move over silent print and *hear* the writer speaking. Voice is the quality in writing, more than any other, that makes the reader read on, that makes the reader interested in what is being said and makes the reader trust the person who is saying it. We return to the columns, articles, poems, books we like because of the writer's individual voice. Voice is the music in language.

Many of the qualities writers call voice have been called style in the past, but writers today generally reject that term. Style seems something can be bought off the rack, something that can be easily imitated. Tone is another word that is used but it seems limited to one aspect of writing. Voice is a more human term, and one with which we are familiar.

We all know—and make use of—the individual quality of voice. We recognize the voice of each member of our family from another room; we recognize the voices of our friends down the dormitory corridor or across the dining hall. And we know that voice isn't just the sound of the voice, it is the way each person says things. We enjoy Anne's stories; Kevin's quick rejoinders; the fascinating details Andrea calls attention to; Mark's quiet, straight-faced humor; Tori's mock anger; Reggie's genuine rage.

Their voices reflect the way they see the world, how they think, how they feel, how they make us pay attention to the world they see. And we are used to using our own voices—plural intended—to tell others our concerns, our demands, our needs.

When we write we should write out loud, hearing what we say as—or just before—we say it. The magic of writing is that the words on the page are heard by the reader. Individual writer *speaks* to individual reader. The heard quality of speech is put into writing by the experienced writer.

Voice also has a political element. Voice speaks out; voice demands to be heard. The person who has voice is empowered. A person whose voice can be heard in writing has an opportunity to influence the policies of a government, a school, an agency, a corporation, a society. Voices that demand listening may attract hearers who may add their voices to the cause. Information is power and voice gives information focus and significance.

Hearing All Your Voices

Voice is often seen as a mystery, an element in writing that is sophisticated, difficult for the student to understand. Baloney. I have never had a student who did not come to the first class knowing—and using—many voices.

Before going to first grade children know there are voices they use in playing they may not use in church, ways of speaking that are not appropriate in front of grandmother, voices that will win permission from one parent and not another, voices that will make peers come over to play or run away home. We all speak before language, when we cry from the crib, with many voices, and it is all those voices that may eventually be turned into voices that will arise from our pages as we write.

Ethnic Influences

We are the product of our racial heritage. There is such a thing as Jewish humor or Black humor—which may not be African-American humor. Sorry, that may be an example of Scottish humor. On my first visit to Scotland I looked up a relative and found an old man digging in a garden. "Are you Donald Bell?" I asked. "Guilty as charged," he answered. I had thought that was a family joke in America and found it was Scottish humor that emigrated to America. It was not a family but a typically Scottish retort. My voice—and yours—is a product of your heritage, all those elements that arrive in your genes.

Regional Influences

Our voices are also the product of the speaking habits of the area in which we are brought up. My speech is urban not country, street language not field language; I speak fast and say "Cuber" for "Cuba" and "Hahvud Squah" for "Harvard Square"; I speak "Boston"—perhaps the ugliest accent in America—not Boston Brahmin but Boston working class. Our ears pick up the patterns of speech around us and we make them our own.

Family Influences

Each family has its way of speaking and we learn speech by imitation. No wonder, that for the rest of our lives, we hear the ghosts of those family members who are dead or live far away, in our spoken and written voices.

Daily Influences

We all swim in a sea of language. We hear language from the radio and television, from the people around us and, by telephone, from those far away. We respond *to* language: We read letters, flyers, brochures, newspapers, magazines, books. We respond *with* language: We write papers, memos, input to computer bulletin boards, notes to ourselves.

Your Language or Mine?

School's view of language often seems authoritative, a matter of rules, as if language were stopped at this moment, a matter of clear right and wrong. School doesn't really mean that. The history of language is the history of change but school has the responsibility to communicate the traditions of the moment, to help students know how language is currently used so people can communicate with each other.

This inevitably gets mixed up with status and etiquette. The "educated" person speaks differently from the "uneducated" one; the person in power uses language differently from the person out of power; the well mannered person speaks differently from the uncouth, uncultured one.

I am uncomfortable with those ideas but they have a truth. I do not come from a well educated family. They did pay a great deal of attention to speaking properly, but my education isolated me from my background and my family. I recognize the need of people to learn the language of those who have power over them, but I also respect the languages and dialects of all the diverse cultures in our society. The grandmother who brought me up spoke Gaelic but she would not teach a word of it to me. I was to be an "American" and to speak "American." Sadly I became monolingual and was cheated of the heritage of Gaelic literature, oral and written, with which I should have been familiar.

I am writing a textbook of revision and inevitably I seem to say "Write like me" and that makes me uncomfortable. And yet, the world judges you, more than it should, by how you speak and write.

If you want to be heard, to be empowered, you have to find your way to use our language, not your own, and eventually to enrich our language with your own.

The Importance of Your Voice

Voice is the most important element in writing. It is what attracts, holds, and persuades your readers.

Significance

Voice illuminates information. Voice makes what appears to be insignificant information significant, and an ineffective voice can make what is significant for readers appear to be insignificant.

Character

Voice is a matter of character. The great essayist E. B. White once said, "Style results more from what a person is than from what he knows." When we write we reveal how we think, how we feel, how we care, how we respond to the world.

Trust

An effective voice demonstrates it is an authority on the subject. It speaks with confidence in specific terms. It is sure enough to qualify, to admit problems, to allow weaknesses of argument; it does not shout and bang the desk but speaks quietly to the individual reader, for writing is a private act—one writer to one reader. The writer's voice endeavors to earn the trust of the reader.

Music

Language is music. Writing is heard as it is read. The effective voice is tuned to the message, the situation, and the reader. The music of the writer's voice, similar to the music accompanying the movie, supports and advances the meaning of the entire text.

Communication

The effective voice can be heard, respected, and understood by a reader. Writing is a public act performed in private and received in private. Both reader and writer are alone. The writer must anticipate the language of the reader so that the act of writing will be completed when the writer's message is absorbed by the reader.

The Expected Voice

Society has voices it expects from us according to the message we have to deliver and the place it will be delivered. The voice of the victorious locker room voice is different from the losing one; the funeral voice different from the party voice. When we read science fiction, a newspaper sports story, a war report in a magazine, an economics—or composition—textbook, we have expectations.

The effective writer knows what the reader expects and decides to write within or against those expectations. But if the writer works against the expectations—using a poem to report on a football game, a narrative as a corporation annual report—the reader must be informed in some way that the writer is aware of going against expectations. The reader must be prepared because the expectations of the reader are always strong.

The Formal Voice

In school and at work, we learn the traditions of the formal voice, the literary research paper, the lab report, the nursing notation, the business memo. These formal traditions are usually rigid for good reason. The doctor scanning the nurse's nighttime notations is not looking for an aesthetic experience or a philosophical essay on pain, but clear, specific information to help modify treatment.

The Informal Voice

What we may not understand is that the informal voice has its own traditions. When we write a thank you note to Aunt Agatha, a humor column in a college newspaper, a note stuck on a door to tell a roommate where we are eating, we also follow traditions. The style may be casual, but it is usually in a casual tradition. Jeans and sneaks may be just as traditional as tux and gown. We need to know the tradition, then try to vary it if the tradition interferes with the delivery of our message.

Genre Voices

Narrative, journalism, drama, poetry, biography and autobiography all have their own traditions that you can discover by asking people in that field to tell you the tradition, where it is published, or by reading in that form. There are many languages of lyric poetry or of jazz, folk or rock lyrics, but each belongs to a tradition that can be defined and described.

The Voice of the Draft

Traditional education focusses on—surprise—the traditional voice; untraditional education focusses on the personal voice. I think they both miss the target. We need to know how to use the conventions of traditional language, and we need to be able to hear the sound of our personal voice; but the focus should be on the voice of the individual draft. Voice, as we have said, is a matter of situation; what is appropriate for one message, in one genre, for one reader, may not be in another.

Listen to the Voice of the Draft

We need to train ourselves to listen to the voice that emerges from the draft. We need to hear how language is being used in this particular case. Of course, what we will hear will be a blend of personal and traditional voices woven together for this particular purpose. That should provide the focus: What is needed here? What is strong and right? What needs to be extended and developed?

We have to be able to work at the console of language, mixing the tracks we hear so they work together to produce a combined voice—the voice of the draft—that will communicate our meaning.

─────────────── ▪ *Writing Exercise* ▪ ───────────────

Choose a personal experience that has affected your life and take ten minutes to describe it in writing. If you work on a computer, turn the screen off; if not, try not to pay attention to how your draft looks. Speak the draft out loud; hear what you are saying as you are saying it; follow the beat, the rhythm, the tone, the melody of what you are saying. Stop after ten minutes and read your draft aloud to hear your voice rise from the page.

CHAPTER 10
RE-WRITE TO EDIT

From the beginning of the book we have shown that re-writing is a thinking activity. At each stage of the revision process the evolving draft reveals the meaning in increasing dimension. We write to discover what we have to say. At the last stage of the process, we edit to refine what we have to say. There is discovery, but the discoveries are usually smaller and the role of the reader becomes increasingly important. We edit more to communicate than to explore.

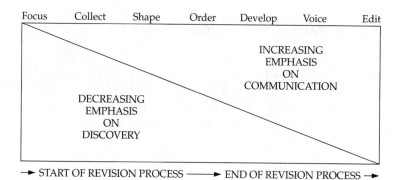

Focus   Collect   Shape   Order   Develop   Voice   Edit

INCREASING
EMPHASIS
ON
COMMUNICATION

DECREASING
EMPHASIS
ON
DISCOVERY

→ START OF REVISION PROCESS ——→ END OF REVISION PROCESS →

■ THE ATTITUDE OF THE EDITING WRITER ■

I used to hate to re-write but my attitude changed with the experience of re-writing: I saw how it improved my drafts. Here are some of the attitudes I bring to the re-writing desk.

Writing Is Re-Writing

It is not an admission of failure when you have to re-write and edit. It is a normal part of the process of making meaning with language. Revision is not a punishment, but an opportunity.

You Are the Reader's Advocate

When you revise, you become the reader's advocate. You step back, drop the natural possessiveness we all have about what we have written and read as a stranger. To get distance on my own copy I often have to name the stranger—an intelligent friend who doesn't care about my subject. He or she will read with disinterest and without familiarity or concern. That's just the reader I want to reach, and so I read my draft through their eyes.

My Ear Revises Better Than My Hand

We spoke before we wrote, historically and individually. Writing is not quite speech written down but it is speech transformed so that it may be heard. The voice lies silent within the page, ready to be turned on by a reader. Writing is an oral / aural act and we do well to edit out loud, hearing the text as we revise and polish it.

The Draft Will Tell You What It Needs

I have learned to respect my draft. Writing is not an ignorant act. Something was happening when the draft was being written. Writers know the contradiction of art: there is usually reason in accident. Try to understand the text on its own terms. Do not make it what you or the world expects, but what the text itself commands.

Welcome Surprise

Many people fear surprise. They hunger for control—and so do I in many parts of my life. But I have trained myself to remember that it is the unexpected that instructs me, the accident, the failure. When I say what I do not expect to say, it is evidence I have been thinking. I have to stop and consider the surprise. It may not mean anything this time, but most times it will mean a great deal. It will point me to my meaning.

Language Is Alive and Changing

This writer sees language as everchanging. I may not like all the changes—I growl at split infinitives and grump when people use host as a verb—but most of the time I delight in our changing language and work as a writer at the edge of tradition. That is where the writer is using language to say what has not quite been said before in a way that has not quite been heard before.

The writer's guide is not right or wrong but what works and what doesn't. No decision about language can be made in the abstract any more than a surgeon should decide to operate without seeing the patient. All editing decisions are context-oriented; what may be correct in one place may not be in another. Writing's job is not to be correct but to communicate meaning.

Accept Limitations

I accept the limitations of my craft—the assigned length of the draft, the expected form and tone, the targeted reader, the deadline—then go beyond the acceptance to view the limitations as a creative challenge. The mural is different from the miniature, the song from the opera, the jazz combo from the big band. The limitations of any art contribute to its breakthroughs; it is not discipline

of freedom alone that are at the center of craft but the tension between freedom and discipline.

Establish Achievable Standards

Student writers and professional writers, myself certainly included,
tend to dream an impossible draft. That is a certain route to failure.
We give up before beginning the draft, knowing we can't do it; we
quit while drafting because we are not living up to an imaginary
standard; we toss the final draft because it doesn't measure up to an
unreasonable standard.

■ THE CRAFT OF EDITING ■

When I write and re-write there is an unconscious element in what
I do. I do not want to become so aware of how my feet are placed as
I go downstairs that I will cross them and tumble down three
flights. I often work instinctively. As I pass through the revision
process, more and more writing is self-conscious. I need to be aware
of how I do what I do, of my reader, of tradition, of my purpose.

The Tools of Revision

Beginning writers have too much respect for their written drafts.
They have been taught to respect—or fear, or stand in awe of, or to
admire without question—the printed text. The writing, especially
if it is typed, appears finished.

The experienced writer likes nothing so much as despoiling a
neatly printed text. The writer cuts right into the neatness, messing
around. The writer cuts and adds and moves around and puts back
what was just cut and discards and re-drafts.

Now I write on a computer where my best friend is the button
marked "DELETE." I draft and revise while always having a neat,
readable text on my screen. But for decades before the computer
age I revised, and my office still has the tools I used—and occa-
sionally still use—to make my final draft look easy, natural, even
spontaneous.

My tools are: a wastebasket, large; scissors; glue (stapler, scotch tape); a black, thick-line felt marker; an extra-fine black pen. I can discard, cut and paste, cross out and insert.

Some of the ways the writer marks an evolving draft include:

Cross-out.	The ~~lazy~~ dog runs.
Take out.	The lazy dog runs/slowly.
Put back in.	The ~~lazy~~ dog runs slowly.
Transpose.	The lazy dog runs slowly
Insert.	The lazy dog ^saunters ~~runs slowly~~.
Move.	The lazy dog runs slowly.
Period.	The lazy dog runs slowly.
Capital.	The lazy dog runs. slowly.

The important thing is to mark the text so that you can see the changes and read the text as it will appear after the changes are made to see if more changes are necessary—and they usually are, because the particular changes the meaning, and the meaning influences the particular.

The Revision Checklist

As we read during revision, moving back and forth from meaning to specific, we deal with a complex blend of overlapping concerns. It is helpful, especially in the beginning, to develop your own checklist of the elements that contribute to voice. Here is my checklist:

Is it specific?

Is it true?

Is it me?

Does it fit?

Is it clear?

Will the reader read?

Does it advance the meaning?

Does it use tradition?

Does it flow?

Is It Specific?

I was surprised to discover this was my first concern. But I find this appropriate after reconsidering it. We write with information, not free floating language; words are the symbols for specific information. And the more specific the language, the more the reader believes and trusts the writer. Specific words also give off vibrations, multiplying their effect on the reader: "dance" is a good noun and it may be the accurate one, but "prom" carries with it special emanations, as does "ball." Finally, specific language is lively, easy and interesting to read.

Is It True?

Unfortunately, it is easy to lie with specifics, and I find myself hyping my own copy. I have to stop and make sure that everything, word-by-word, is accurate in itself and in the context of the line, the paragraph, the section, the entire work.

Is It Me?

It has taken me most of my lifetime to accept myself. I have wanted to be old when I was young, young when I became old; heavy when I was a string bean, thin when I became a pear. I have to tune my voice to my own way of making music on the page.

Does It Fit?

Nothing in writing stands by itself; we have to be master cabinet makers who can fit everything together so that our final drafts contain their full load of meaning and stand up to a reader's scrutiny. The word has to fit the words before it and the ones that come next; the phrase has to fit the line; the line the paragraph; the paragraph the section; the section the whole piece.

Is It Clear?

When I am revising my draft, I try to practice George Orwell's rule: *"Good writing is like a window pane."* I do not want to call attention to myself but to my subject; I do not want to get between the reader and the subject.

Will the Reader Read?

I am my first reader and I have to make myself the attorney for all the readers that will follow. Is the information placed in a position that gives it proper emphasis; have I been as simple as possible without oversimplifying the subject; is the draft paced so that the reader is propelled forward but still allowed time to absorb each point before being moved on to the next one; have I asked the reader's questions—and answered them?

Does It Advance the Meaning?

Every piece of information, every word and the sound of every word, every comma and semi-colon, every sound, every space, should serve the meaning of the draft. No matter how interesting an anecdote you have, how dramatic a quote, how fascinating a fact, how clever a phrase, it should not survive revision if it does not advance the meaning of the draft in the reader's mind.

Does It Use Tradition?

The traditions of our written culture—rhetoric, grammar, mechanics, usage, spelling—are the record of the conventions that successful writers in the past and present have used to communicate meaning. Conform to them when they help you communicate your meaning, as they will most of the time, and go against expectation when they do not.

Does It Flow?

After all this fiddling around, cutting that, inserting this, moving the other stuff around, making changes in information and position, meaning and sound, the draft must be smoothed over so that none of the effort shows. The music of the final draft should flow with an easy naturalness so the reader is absorbed in what is said, not how the writer has said it.

■ REVISING THE WORD ■

The craft of writing—and the art—begins with the word that ignites the draft, and it ends with the word that will rise off the page and enter the reader's brain. Not any word: the right word. Mark Twain said it best: "The difference between the right word and the almost-right word is the difference between lightning and a lightning-bug."

A word has no value unless it communicates information, not the fancy word, not the word to impress, but the simple word that carries meaning directly and clearly to the reader's mind.

Apply the checklist on pages 187-188 to the words in your draft.

Notes for the Writer Revising the Word

I am going to take you inside the craft of revision with one writer—myself—and share the mental notes I keep in my mind as I revise the word.

Find the Right Word

The game of words is one of the most fascinating that writers play. We all know a simple word such as "walk." But is it so simple? Walk means simply to put one foot in front of another, but when a woman is asked out to dinner and she answers, "Take a walk" it is not an invitation to romance; the union leader tells the boss "We walk" and the workers go on strike. And think of all the walks there are: stroll, saunter, march, stride, shuffle, promenade, tiptoe, and on and on and on. My job is to find the right word, the single word that reflects the meaning I need to communicate.

There are a number of ways to check on the meaning of the word you have used to see if you have chosen the right word—the word that communicates meaning accurately and gracefully.

Recall all the meanings of the word to see if you have the right one. Most words have many meanings. "Hit," for example, can mean a blow to someone, an attempt at seduction, a gangland murder, a baseball activity, a popular song—and you should be sure you are using the one your reader will understand.

Read *aloud* to hear the meaning of the word. We are most experienced with spoken language, and we will hear a wrong word that we read right over silently.

Consider the *context* in which the word is being used. If I say "I don't know no grammar," *no* is the wrong word but if I am demonstrating my ignorance, it may be the right word *in context*.

Put a check mark beside the words in bold face, capital letters or underlining on the first draft so they stand out when you come to edit the final draft. While writing an early draft you will often use a word you knew was not right but you did not want to slow the flow of the writing to use the dictionary. Sometimes I will put a question mark in parenthesis (?) or brackets [?] after the word or even bracket a note to myself: [cliché], [weak word—make stronger], [vague—make specific].

Look up the definitions of any word you do not use naturally in normal speech—and some common words if you are using them in an unusual way.

Use Simple Words

I am fascinated with the power of simple words. I use the simplest word I can find that carries a full load of appropriate meaning to the reader.

Hear the Music of the Word

Each word performs its own music. "Zip" is different from "zap," "maim" from "hurt," "guffaw" from "giggle," "mumble" from "grunt." I want the music of each word—the musical note—to contribute to the music that supports and communicates the meaning of the draft.

Select the Reader's Word

If our language is different from our reader's language, then we exclude. I want to include. I do not want to write in formal, academic rhetoric, but neither can I speak undergraduate. It is a long time since I was an undergraduate, and if I try to use today's slang I'll miss or sound patronizing, so I try to write in a lively way that draws the reader in.

Define the Word

When I use a word that is unfamiliar to my audience—*"lede"*—I try to define it immediately—*"lede, the newspaper term for the beginning sentences or paragraphs in a news story."* I don't often need to use a footnote or a formal definition and I first try to define the word by the context in which it is used. *"The news writer first concentrates on the beginning of the story, trying to get the lede right before going on."*

Avoid Clichés

Clichés are worn out expressions that belong in the dumpster: "couch potato," "real cool," "freaked out."

Our first draft writing is often full of clichés that were once, like the new TV commercial, amusing or insightful—perhaps both—but are worn out with use (such as "couch potato"): *trite,* so tired I could die; *stereotyped,* cross as a bear; *slang,* airhead; all grown meaningless with overuse.

How do you recognize them? By their familiarity. Listen to the speech of your friends. We all have difficulty hearing our own. Notice how many times you know what people are going to say before they say it. Notice how many times the same phrase is used—"really"—those are clichés. Cut them out. Say what you mean using your own eyes, ears, voice.

Resist Jargon

Every trade has its jargon—"software" may be an example of jargon that has made it into the language because it has a specific meaning—but most jargon is private language that excludes. It is used to demonstrate you are a member of a club and the reader is not. Again, I want to include as many readers as possible in my writing.

And jargon gets sloppy; the same word means one thing to a clinical psychologist, quite another to an experimental psychologist. One of the problems for those of us who use different computer programs is that the jargon of one program is not consistent with another. Avoid jargon. Write in your own language.

Write with Verbs and Nouns

Effective prose is built with nouns and verbs.

I always feel a bit guilty—not the full weight of SIN but a little twinge of guilt—when I use an adverb or an adjective. I use them but it means that I have not found the right verb or noun, that I hope to catch the meaning between two words, and meaning is better caught with one word. At the most important points in my text I double my concentration on finding *the* verb, *the* noun.

Write in the Active Voice

When I was the English Department chairperson, and a faculty member was voted tenure, my note often went something like this: "You got it. Congratulations." When the opposite was true, I was more likely to sound like this: "It was the opinion of the committee, that, after consideration, it would seem appropriate that we would not grant tenure at this time." Passive, wordy and a tone that avoided personal responsibility.

Avoid the passive voice. "John hit Jim" is far more direct—and forty percent shorter—than "Jim was hit by John."

Watch Out for the Verb "To Be"

The verb "to be" in all its forms is a workhorse of our language but we can overwork it: "I will be seeing you outside" may better be "I'll see you outside." Make sure that the verb "to be" doesn't clutter up your draft.

Cut Unnecessary Words

Each word should carry its weight of meaning. This doesn't mean you write like a telegram, but it does mean that you check each word during revision. You may have stuck in such a clutter word as "however." It may need to be cut; but it may also give an important signal that a significant qualification is coming to the reader. Or, it may simply slow down the text so the reader can have a moment to absorb what has been said before receiving new information. Make sure every word is necessary to communicate your meaning. If it isn't: delete.

Some Personal Tips

Make your own list of things to watch out for. Each item may do the job when it is needed, but certain words seem to multiply on their own, spreading throughout a draft. Weed them out. My list:

-ing	—Can I take it out and make my writing more direct?
-ly	—Is this adverb necessary?
that	—Sometimes it seems as if each sentence has a "that" and some have two or three. I cut every one I can.
would	—Most of my "woulds" add nothing to my text. I subtract them.
very	—One of the amazing things about our language is that "very" is usually less rather than more. "It was a very beautiful sunset" is not as beautiful as "It was a beautiful sunset."
quite	—One editor claimed I had at least one "quite" on every page of a book manuscript. She was right. They were zapped.

Make your own list and act on it while tuning your draft.

Spel Koreclie

Researchers promise me that the ability to spell correctly is *not* related to intelligence. I am grateful. I am a poor speller, and so are many writers. Unfortunately, teachers, bosses, editors, readers, do equate intelligence with spelling. When they see one word misspelled they begin to sneer at the writer. I hate sneers. To avoid them I have several solutions:

1. I have a spell checker on my computer and, boy, do I use it. But it has some problems. The computer is dumb. If a word is a correctly spelled word, it does not care if it is the right word. Examples: "to" and "too," "their" and "there." And

on top of that, my poor spelling combined with my poor typing produces words the spell checker cannot find.

2. Beside my typewriter I have *The Word Book.* Mine is published by Houghton Mifflin, but there are similar books on the market. Mine simply has the forty thousand most frequently used words in our language listed alphabetically. I use that every day I revise. If that doesn't work, I go to my large dictionary.

3. If I can't find the word—"pneumonia" should start with an "n" but it doesn't—I go to *The Bad Speller's Dictionary.* Mine is published by Random House, but other publishers have their own versions. It has the wrong spelling followed by the correct one: "neumonia _____ pneumonia."

4. I have a list and you should have your own.

lead, led

its, it's

there, they're, their

to, too

your, you're

lose, loose

occurred

than, then

affect, effect

principle, principal

can not, cannot

And I find that when I'm writing quickly I can hear one word and write down a sound-alike word, a homonym: "altar" instead of "alter," "complement" instead of "compliment," "conscience" for "conscious," "desert" for "dessert," "hear" for "here," "know for no," and I have to clear those howlers out of my final draft.

And rules may help: *Lice:* "i" before "e" except after "c." Believe. Receive.

The important thing is to make your own list and prop it up in front of you when you tune your final draft.

The Dictionary

The dictionary is as essential a tool to the writer as the hammer is to the carpenter. In the dictionary the meanings of words are recorded. The dictionary is the referee that makes the call, most of the time, as to whether the writer is using the right word to communicate meaning.

The Thesaurus

And what about that thing called a thesaurus and credited to Mr. Roget? It is a popular gift to a student going to college and most word processing computer programs even include a thesaurus. If you do use a thesaurus, be sure to look up the meaning of the word you choose.

■ REVISING THE LINE ■

After the word comes the line, that unit of writing that may—or may not—become a sentence. The line is that unit of words that surround *the* word, that set off and display the word, and in which two words can ignite to create a meaning that is beyond either of them alone.

Apply the checklist on pages 187-188 to your draft, concentrating on the line.

Forms of the Line

The effective writer has mastered the line in all of its forms. It is the line that provides the energy for each piece of writing; it propels the material and the reader forward.

The Phrase

The phrase is the unit of words that connects the single word to other words so that meaning is created. It always confused me in school that we studied vocabulary, then leapt to the sentence. I usually wrote in small units of words as much as with a single word. I now know I was right. We write with phrases, those small units of language in which words collide and give off a meaning that is different from either word alone. In writing, two and two can make seven or eleven.

The writer is forever excited by what happens when words interact with one another, igniting a thought, a feeling, a vision, an insight in the mind of the reader. Perhaps because this New Englander experienced a flash flood in New Mexico, I have forever been fascinated by the energy given off by those two words: "flash flood." They create an accurate and terrifying picture and their music underscores the meaning. Say the words aloud and hear their hard edged quickness. They sound like a flash flood.

At the center of language is the phrase that captures an idea or a mood, but there are some dangers the writer has to guard against. The phrase has so much power that politicians and advertising agencies have become skillful at the glib phrase that sounds good and says little. Our presidential campaigns are no longer debates in which issues are explored but "sound bites" (an interesting phrase in itself), skirmishes in which fragments of language are hurled between candidates. Apart from the issue of abortion, no one can deny the advantage anti-abortion forces gained when they established themselves as being "pro-life." Their opponents, not willing to be "pro-death" rushed to establish their position as being "pro-choice." These phrases oversimplify a complex human/political/theological/medical issue.

Pay attention to the way words rub against each other. When two words strike each other they often give off a meaning that is unexpected and powerful: "She had a whim of steel."

The Fragment

I see the fragment as the first cousin to the phrase. Formal instruction in writing commands: No sentence fragments. But today's writers qualify that edict: Don't use sentence fragments that don't work. (Watch out for double negatives: Use sentence fragments that work.)

A fragment is a non-sentence: "Tom and John in pickup. Gone fishing." To be a sentence, it should read, "Tom and John hopped in the pickup to go fishing." *But* if the previous sentence read, "The private investigator watched, then scribbled in his notebook, 'Tom and John in pick up. Gone fishing.'" then the fragment would be correct. If you write a sentence fragment and it must remain that way, keep it, but 996 times out of a thousand it will work better as a sentence.

The Clause

The clause is a phrase used within a sentence, usually set off by commas, to expand or qualify the meaning in a sentence. How many clauses should there be in a sentence? Just enough. Enough for what? To communicate the meaning of the sentence.

The Sentence

The sentence communicates meaning. That subject–verb–object sentence demonstrates the full glory of the line. The subject–verb sentence is the fundamental force in written—and spoken—language. To write English well, master that simple, direct sentence and then build it up with carefully crafted clauses that contain the ideas that cannot be contained in the simple sentence.

Do not, however, fall into the trap of thinking that the more complex the sentence, the more impressive the writer. First, be a master of the subject-verb sentence and return to it when you communicate your most important ideas. Respect its clarity, simplicity, energy, grace.

Editing the Line

Now I am going to invite you to visit my workroom and allow you to hear the messages I mutter and mumble to myself as I edit the line.

How Many Words Should There Be in a Line?

Well, if I hear correctly, paying attention to the arts of communication as I understand them, and, of course, to the parameters of language usage, it is not, however, just a matter of correctness, but has socio-economic components as well, that I would advise the writer to cut what can be cut.

Cut what can be cut.

Cutting is one of the great satisfactions of revision. Eliminate the unnecessary and the necessary runs free.

Short or Long?

Short for emphasis—"Jesus wept"—with the longer sentences in between. That's my first rule. Let the sentence roll along when that is appropriate to the idea; allow what is being said to determine how it is being said.

And I need to remember to vary my sentence length and design so that I create an attractive—and appropriate—pattern of prose that unrolls before the reader's eye and ear in an interesting fashion.

Finding the Point of Emphasis

The greatest point of emphasis is usually at the end of the line: the next at the beginning. Move the most important piece of information you want to communicate around within the sentence until you hear where it has the greatest impact.

Writing with Metaphors

Robert Frost said, "Poetry is metaphor, saying one thing and meaning another, saying one thing in terms of another." And so is all lively writing, prose or poetry. "He is a nut," "the exam bulldozed me;" "we had a ball." This is the pure magic in language. By metaphor we liven, illuminate, communicate; the reader reads a metaphor and sees, in an instant, what we mean, how we feel.

Parallelism

When I arrange words so they reveal, I want my reader to build toward a conclusion that will seem inevitable. Therefore, the pattern of my words, in this case my clauses, must work in parallel, repeating the pattern that I have established in the beginning. Obviously, I need to line up these clauses so they are similar and make my meaning clear to the reader.

Listen for the Beat

The line, in all it forms, has a beat. It is not a metronome beat, an even regularity, but a beat that reflects the meaning of the draft. Listen to that beat; then play to it with the choice of words, the length of the line, the pattern of emphasis in the line.

What Comes Before; What Follows

Writing the line, I have developed triple vision. I am aware of what I am writing, what came just before, what may be written next. The line grows out of what I have written and grows into what will be written next.

Transitions

Some editors and teachers like transitions—meanwhile back at the ranch—but I try not to write them. Most formal transitions mean that I do not have the material where the reader needs it. If I manage to answer the reader's questions *when they are asked,* I will not have to write transitional phrases.

Pass the Machete, Please

One of the great satisfactions of revision is cutting through the twisted tangle of lines that wind back on themselves, hiding the intended meaning and blocking the way of the reader. Some of the solutions:

- Cut what can be cut.
- Try one simple subject-verb-object sentence.
- Try two or three subject-verb-object sentences.
- Make it more specific.
- Use simpler words and *active* verbs.
- Try a list. (As I have done here.)

Pronouns

Make sure the "he" relates to a "he," the "she" to a "she," "it" to an "it," and make sure they send the reader back to the right person or thing. It's easy to slip while writing and easy to fix while revising.

Sexist Language

If we say "he" meaning everyone who is a writer, a diplomat, a teacher, an athlete, we exclude the majority sex. If we switch all "he's" to "she's," we attempt to correct the sins of the past but create new ones. It is easy to get rid of sexist—and racist—language if you eliminate any words that are offensive to readers of a particular gender or background and if you use plurals or specific names so that you do not have to attempt the generic "he."

Ifs, Ands, and Buts

And it is all right to use "and," "if," "but" to start a sentence or paragraph if that is the most effective way to do it. Don't fall into a repetitive pattern of lines; use different ways of writing a sentence, making sure each one serves a reader.

Starting with a Dependent Clause

While trying to make a complex idea clear, it is important to write sentences that run clear and fast. The reader has to hold that first clause that is dependent on what follows until the rest of the sentence is read; then snap it into place. Sometimes it is appropriate to begin a sentence with a dependent clause for emphasis, transition or a change of pace, but watch it. Here's the first sentence of this paragraph without the dependent clause: "It is important to write sentences that run clear and fast while trying to make a complex idea clear." Which sentence delivers the message most clearly? Which one delays, asking you to hold the clause in memory, then read it again after you have finished the entire sentence? Which sentence is most direct?

Maintaining the Tense

In revising, I notice that I say I am going to do something, and in the same line I used the past tense "said." Make sure that your tenses agree. To what? To each other and to the draft's relationship to what is being reported. I like to write in the present tense whenever possible because it is lively and immediate, but I can only do it when what I am writing about makes it possible.

Leaving the Reader Dangling

Don't. The reader doesn't like to dangle. It's uncomfortable and confusing. Writing this book, the editor told me to be clear. No, the editor didn't write this book, but that's what I wrote by dangling.

Running On and On and On

"It sure do, don't it." When I read my own sentences and start to run out of breath, I chop them up and give my readers a chance to breathe.

Punctuation

I once had a writing professor, Carroll Towle, who said you use commas when the reader has to breathe. That helped, but it is a little too simple. Brock Dethier tells his corporate executive students: "A comma says 'take a small breath'; a semicolon says 'another independent clause is coming'; a colon says 'here comes a list, definition, or explanation.'"

Punctuation is largely a matter of meaning; punctuate to make your meaning clear.

Dashes, Brackets, Parentheses

The dash has been frowned on by grammarians—if you are writing for a grammarian, don't use it—but I find it is a wonderful device to use to interrupt your line to comment on what you have just written, to give information the reader needs at the moment, to qualify and emphasize. And I delight in dash, bracket and parentheses.

Be Spontaneous

John Kenneth Galbraith, an economist who is known for his lively writing on complex economic issues once said:

> *In my own case there are days when the result is so bad that no fewer than five revisions are required. However, when I'm greatly inspired, only four revisions are needed before, as I've often said, I put in that note of spontaneity which even my meanest critics concede.*

I am most spontaneous while revising the line. Now that I have my meaning I can make it dance, grump, growl, preach, or just be simple and clear—whatever is appropriate to my subject. One of the greatest joys in writing is revising the line, becoming, through revision, fresh and spontaneous.

■ REVISING THE PARAGRAPH ■

I owe the late Hannah Lees an enormous debt. We had the same agent, and when I met her and discussed the problems I was having with my first magazine articles, she suggested I cut my typewriter paper in half and write one paragraph to a page. I followed her counsel. I began to learn how to craft the paragraph, the basic rhetorical unit that carries meaning to the reader. This is, I believe, my twentieth book, but a book is nothing but a long series of paragraphs; the book is built, like a brick wall, a paragraph at a time.

I revise by making sure, paragraph-by-paragraph, that I am delivering meaning to my reader in a voice the reader will hear and understand.

The Extended Paragraph

I wanted to call this section "the chunk" because the term "paragraph" seemed too restrictive. Some of my "paragraphs" are verses of poetry; some paragraphs in my fiction—and even in my non-fiction—are small scenes with dialogue, action and reaction, setting; others are anecdotes, the small narratives so common in popular non-fiction. As you read this section, realize that sometimes the paragraph is only one sentence long. At other times it grows beyond the typical paragraph into a longer unit designed to carry meaning to the reader.

Apply the checklist on pages 187-188 to all forms of the paragraph in your draft.

Rereading the Paragraph

Here are some of the thoughts that run through my mind as I fine tune my paragraphs.

The Topic Sentence

Avoid it. Each paragraph has to have a topic, but most paragraphs should not announce it. That insults the reader and slows down the text.

Long Paragraphs or Short

Most of my paragraphs are short.

This does not delight everybody. Many academics feel my paragraphs are too short, and some feel that's true of my words and my sentences. My first textbook was turned down by one publisher because "It will be read for enjoyment, not instruction." I am charged with writing "journalistic" paragraphs, and I was certainly influenced by writing for newspapers, where five lines of type-written prose becomes ten lines in print. I can and do write long paragraphs in my novels, articles and textbooks when I need the space to develop my ideas, when I think the audience will be inter-ested enough to read long paragraphs, and when I know the text will fill a full page.

I deliver small loads of information and develop my ideas over a series of paragraphs when other writers would develop the same material in one paragraph. The goal, however, is not to follow a specific paragraph length requirement but to develop and deliver information and ideas in a well-tuned voice.

You should not write in one-length paragraphs but in a pleasing and effective variety of paragraph lengths. Normally, you should use shorter paragraphs for emphasis or clarity of complex ideas. Longer paragraphs carry the reader to and away from those peaks of meaning.

And if your editor, teacher, or the design of the publication in which you are published requires longer or shorter paragraphs, de-liver them.

Developing the Paragraph

The old-fashioned instruction, CUE—coherence, unity, and em-phasis, is not often taught these days, but it is one high school in-struction I remember and try to follow.

Unity	Everything in the paragraph should be about one thing.
Coherence	One thing should logically lead to the next.
Emphasis	The main point of the paragraph should be clear.

Attribution

The reader should know where significant, controversial, or unexpected information is coming from. This can be done in many ways. It may be obvious from the text above; it may be gracefully woven into the text; it may be delivered in an announcing or attribution line; it may be in a footnote. You should try, however, to do it in such a way that it does not draw the reader's attention away from the information the reader needs.

Opening Paragraphs

The most important paragraph in any draft is the first one. It attracts the reader, establishes the subject, the tone, the form, the length and pace of the draft. It is vital to fine tune the opening paragraph that sets the standards for all the paragraphs that follow.

Descriptive Paragraphs

Description—and I mean description of an idea, theory, process as well as a physical description—is a major function of paragraph crafting. It is important to work from a single angle of vision and show everything in relation to the dominant impression created by the angle of vision. I also like to use the word "reveal," to remind me to allow the subject to reveal itself in a way natural to the subject.

Dialogue

Dialogue is easy. It is action. "Get thee to a nunnery." When Hamlet says that to Ophelia, he is committing an action against her, and she has to react by going to a nunnery or saying, "Shove off, Hamlet. I'm going to a rock concert with Boris."

Then he answers, "Not, Boris."

And she says, "Yes, Boris and I intend. . . . "

Dialogue is constructed from the interaction of characters. Usually, each person's speech is a separate line or paragraph. Dialogue is lively, revealing, fun to write and fun to read. It shows more than it tells.

Closing Paragraphs

Almost as important as the opening ones. This is what the reader will most remember. In general, I try not to tell the reader what to think or feel in the last paragraph, but to give the reader information that will make the reader think or feel.

■ USING TYPOGRAPHY TO MAKE MEANING CLEAR ■

The computer and the development of desktop publishing has multiplied the typographical options available to the writer. For example:

The way the text looks can dramatize its meaning.

The way the text looks can dramatize its meaning.

The way the text looks can dramatize its meaning.

The way the text looks can dramatize its meaning.

The way the text looks can dramatize its meaning.

~~The way the text looks can dramatize its meaning.~~

The way the text looks can dramatize its meaning.

The way the text looks can . . .

The writer can also box part of the text, use borders, typographical devices (○ ■ ¿ ☎ ✉ ✄ ♡ ♀ ♂ ★) to emphasize the message of the text.

The danger is that the devices will call attention to themselves and not to what is said, so a good rule is to use them sparingly. To make your meaning clear and dramatic you will want to:

Clarify

Break up the text with major headings. These usually indicate a new section by being centered on the page. Identify units of meaning within these major sections with sub-heads in the margin. Further indented sub-heads can indicate even smaller units of meaning. Usually the heads show their importance through a coherent and consistent system of capital letters and small letters; underlining, bold face, and italics. Sometimes the sections are numbered with Roman numerals, Arabic numbers, and letters which allow the writer's outline to show and help the reader understand the organization of complex material and to be able to refer to it efficiently.

Emphasize

It is important that the reader know your central message and the material on which it is built. This can be done with an indented list of the main points, often single-spaced if the text is double-spaced. The writer may also use boldface, underlining and capital letters to call attention to the most significant parts of the text.

Visualize

With the increase in the use of sophisticated computers, more and more writers are able to clarify and emphasize with graphs, charts, and increasingly clever forms of graphic communication.

■ EDITING THE FINAL DRAFT ■

Company's coming! Run through and put the dishes in the clothes hamper, the laundry in the trunk of the car; make the bed; swish out the shower; fluff up the pillows on the sofa. Now, at the end of the revision process, read through the draft quickly using the checklist on pages 187-188 to make the copy as clean and readable as possible.

Take a draft and read through it four times. Tedious? It shouldn't be because you will be rewarded with little surprises of insight and understanding along the way. Apprehensive? Scared? Afraid you'll make a mistake? Don't be. You aren't so much trying to make the draft correct as much as tuning your voice so it will rise clear and strong from the page.

■ EDITING ACADEMIC WRITING ■

Remember, an academic paper is a demonstration of disciplined thought. Here is a checklist for editing the final draft of academic writing.

- *Is this house of meaning built on a solid foundation of accurate information?*

- *Are the sources of your information clear to the reader?* An academic paper is written for members of an academic community who are doing their own research, scholarship, thinking, on the topic of your paper. They need to see the source of your information not only to check its veracity but to explore those sources themselves as they investigate the topic.

- *Does your text take a critical view of the information?* Academic writing is not the accumulation of information but the product of an informed intelligence that has thought about the importance of the information. "Critical" does not mean "negative," but it does mean having an opinion that can be supported with specific evidence.

- *Is the information in a context so its significance is clear to the reader?* The academic writer must include the environment that makes the writer's opinion worth paying attention to. That context must be clear to the reader who is inclined to snarl, "So what?" The effective academic answers the "So what?"

- *Is the meaning of the text developed in a logical, objective manner?* The reader of an academic paper expects to see the writer's thinking in process. The reader wants to examine the construction of the writer's thesis, step-by-step, each point growing out of the last point.

- *Are the reader's questions anticipated and answered?* All readers are engaged in a dialogue with the writers they read but this is especially true of the members of the academic community who read actively, engaging in a process of intellectual interaction with the writer.

- *Is the voice appropriate to the topic and its readers?* Voice, the music of discourse, is usually more detached than personal writing but it varies according to the topic and the audience. Still, an academic voice is a writer's voice tuned to the purpose of the text and its potential readers. Voice reveals the person behind the text and the sense of an individual writer is as important in academic writing as it is in any genre.

- *Is the final draft clear; does it break the conventions of writing only when it increases the clarity and grace of the text?* Good writing is good writing in any genre. Academic writing should have the energy, grace, humor, clarity—above all, clarity—of writing in any form. Nothing should get between the mind of the reader and the mind of the writer as they—together—confront the topic.

STUDENT CASE HISTORY

Roger LePage, Jr.

It is the ultimate compliment when I work through a young writer's paper line-by-line. Many students want this attention but there are reasons not to do it most of the time.

One, you can't do an effective job of line-by-line editing of your own draft—or anyone else's—unless the writer has something to say that is worth saying. As Wallace Stegner said, "You can't sharpen a knife on a wheel of cheese."

Two, I do not know the writer's subject, and my editing must be based on understanding what the student has written—and what the student has known but left out. Most writing is left out, in the sense that good writing grows from abundant soil. I am fearful I will take the piece of writing away from the student as I edit it, inevitably to *my* vision of the student's world. For example, Roger LePage's fine piece below describes the Rollinsford, New Hampshire dump. But it is also a story of growing up that will cause readers to relive part of their childhood and, in so doing, understand themselves better than they have before. Good writers can make readers laugh and care and remember and think—LePage does all of that and more.

I live nearby but have never been to that dump. I do not have the specific details in my notebook and my memory from which effective writing is constructed. I do not have Roger's vision of this world and I do not have his way of using our language. There is a conflict here: I want Roger to see one professional writer at work on his prose, but I fear I will appropriate his prose and make it mine.

And when I do edit line-by-line, I am always fearful the student will take me too seriously. I only tend to do this with the best, most resistant students. The student who slavishly follows my editing will learn nothing; the student who questions what I have done and how I have done it may learn a great deal.

The Student's Original Draft

Summer days in Rollinsford, New Hampshire were very boring until we discovered the dump. As ten year olds we didn't have summer jobs—except, of course, finding amusement and adventure. We didn't even do that well until the time one of us, maybe me, maybe one of the others, steered his bike off of Main Street and led the pack onto the dirt dump road. The intriguing pillar of smoke that was perpetually rising above the trees from the dump may have drawn us in. Or it could have been the "NO TRESPASSING" sign and the locked gate—things like that always invited us into abandoned houses. Whatever the reason; we discovered the dump and were not bored for the rest of the summer.

That first day we didn't even make it into the actual dump. The dirt road leading in was about half mile long, and about halfway down we discovered a slimy little pond with lots of frogs and snakes around its fringe. We spent most of the day hunting them and trying to catch them. This didn't hold its appeal too long for me though. One of my friends, Jason, was very cruel and whenever he caught a frog he would first put it on the road and torment it with a stick, and then he would pick it up and squeeze it to death in his hand and throw the guts at whoever was in range. I was made target twice and decided it was time to go home.

But soon we made it past all the distractions along the road, which were many, including the rusting corpses of several kitchen appliances and one time there was the ass end of a deer the stench of which was still strong for a quarter mile radius. But nothing could be compared to the adventures found in the dump itself. We began in the huge pile of discarded tires, immediately abandoning our bikes right in front of it and trying to race to the top. "King of Tire Mountain" lasted for weeks, producing many bruises and scuffles and many hurt feelings. Then we built forts within the inner depths of Tire Mountain—a much more peaceful game. When it was time to try to furnish our forts we began exploring the junk in the rest of the dump.

That was when we discovered the greatest thing possible in a ten year old boy's mind: dirty magazines. I'm not sure who discovered them first, but there was soon a huge uproar and race to see who could collect and hoard the most. We would bring them back to our forts and pile them up, going through each one with great enthusiasm. We would organize them according to their appeal to us: the ones with girls showing everything spread out and accessible were our favorites, the ones with boys and girls were second, the plain boob shots were last. We had another pile of 'girls with girls' which we didn't know how to take but spent the most time looking at. We would move around in funny ways, always moving around to maximize that great new ache in our pants while looking at the magazines, and we all seemed to have the shakes.

* * * *

I would also go to the dump on Saturday mornings with my dad. It was part of our Saturday morning errands: the dump, the barbers, the grocery store, then church. Most of the time he didn't even have anything to throw away. It was a Saturday morning meeting place for the true Rollinsford men—those who still called the town Salmon Falls, and a group to which my dad belongs because he was born and raised there. It was also a place of refuge from their wives: Friday night was poker night, (I could always hear the door slam from my bedroom when my dad stumbled in just before day-break,) and on Saturday morning the men were red-eyed and 'in the dog house' as they put it.

But my dad would go to the dump mainly to pay his respects to the toothless old Greek, who I've only heard referred to as The Greek, and who was a great drinking partner of my dad's father. The Greek worked there. He made sure that you didn't put metals or combustibles in with the burnables and he took his job very seriously. He tended the fire and if you tried to interfere, his toothless smile would disappear and he would straightway banish you. Banishment from the dump is one of the lowest dishonors in our town.

You wouldn't think anyone would be overly enthusiastic about being friends with the man who tends the dump, but in towns like mine occupation is not nearly as socially relevant as age or drinking ability. There were two other reasons why The Greek was respected: he did once have an occupation, but he retired and instead of sitting around all day, or bagging groceries in some supermarket, he opted to tend the dump. That choice was respectable and The Greek knew this and took his job very seriously. Also, he was a great old drinker and fighter, actually legendary. His missing teeth were the proof. Everybody knew that 'he got them knocked out' in a fight years back. The important thing was that 'he got them knocked out,' not 'somebody knocked his teeth out' because the latter would imply that The Greek was not in control at the time. And even though it was rumored that his wife, also something of a drinker and fighter herself, has his tooth marks on her knuckles, The Greek was still respected;

he earned the right to be toothless. He brandished this honor too, by smiling all the time—but I never thought The Greek was such a happy man.

After my friends and I started spending so much time in the dump I began to look at the Saturday morning visits with my dad in a new way. I fancied myself a spy, watching our forts as inconspicuously as possible and always listening for clues to find out if anyone was on to us. I thought that sometimes The Greek would catch me looking at the tire pile and possibly know our secret, but I wasn't sure. I felt that it was just strange that he looked at me at all, because before my friends and I started playing at the dump The Greek never seemed to notice me. I thought then, that maybe I just never noticed him noticing me before because it wouldn't have mattered if he did. Whatever the case, I was wary.

* * * *

Then one day we were playing at the dump, alternating between culling our dirty magazines and our next best discovery: combustibles. It was Jason, the sadist frog squisher who discovered this one, I'm sure, he was fond of destruction in any form. On one of his devilish whims, he threw a paint can into the dump fire and silently waited for something to happen. The rest of us were busy going through junk and exploring and didn't even notice. The explosion sent us running for cover, yelling all of the profanities we knew (most of which we got out of our new magazine collection,) and nearly pissing our pants. When we realized what had taken place, the excitement of the whole thing brought us together giggling, wrestling and rolling around in the dirt. This was probably why we didn't hear the pick-up coming down the dump road.

When we did finally notice it, it had pulled up right beside us. We all jumped to our feet and stood there, not sure what to do. I recognized the green, rusty, beat up old thing—it was The Greek's truck. He got out and stared at us with full blown adult contempt, "What are you kids doing? You the ones been down here vandalizing?"

Even at ten I wondered how the hell anybody could vandalize a dump, but as I said, The Greek took his job very seriously.

"Aren't you Roger's boy? I knew I'd catch you down here," he stared at me with his exaggerated toothless frown and shook his head. His nose looked like a spoiled piece of fruit. "There's nothing but piss and vinegar running through your veins. I know. Knew your grandfather, same thing. Your dad too," he was still staring and shaking his head, the loose flaps of his jaw skin shook around violently. "I knew I'd catch you down here. How would you like me to tell your old man what you've been doing?"

No ten year old wants his parents to know what he does on summer afternoons; it's a child's first glimpse of independence and the privacy is treasured, locked away in the child's heart. When one comes along and tries to adulterate that treasure, as adults always do, the loss of the treasure is a far greater tragedy than any thought of impending punishment. I asked him meekly, "Please don't tell."

The Greek looked at me and then at my companions. He turned and walked along the dirt towards Tire Mountain. I thought I was going to be sick. The Greek knew! He was on to us. First we would be banished from the dump, then from our homes and then we would be kicked out of the town altogether. Maybe even sent to prison or reform school.

"How would your mothers' feel if they knew you were looking at those kinds of things?"

Oh God! I thought. It's one thing to lose your cherished summer freedom, or even to be locked away, but to have your mother think you're a pervert, to have her think you're one of those weirdos like fat, old Slow Jimmy who smiles at you funny and always asks you to go for a walk. To have your mother know you look at dirty magazines! It was absolutely the worst thing imaginable.

The Greek paused and looked around. He looked at me hard. I'm sure he could see I was about to commit the unspeakable in front of my friends—that I was about to cry. He had done his job, and done it well. The Greek took his job very seriously. "Well, if I catch any of you kids here again, your parents will all get a call from me. And if I catch you here again LePage, I'll give you a beating myself."

So we pedaled the hell out of there, still shaken and guilty. We resolved never to return to the dump, and we didn't . . . at least for the rest of the week. But after all, as kids we also took our jobs very seriously.

A Professional's Editing

I find this a good rite-of-passage piece about an interesting and mysterious place that has depth and texture: children setting out in the world, beginning to have a secret life apart from parents; a son's relationship to his father; the boy's tense relationship with Jason, the cruel leader of the boys; the narrative of their explorations of the world; The Greek, his history and status and what it means. And the writing is good, full of potential as well as wordiness and the un-evenness of language expected from a beginning writer.

I thought he was ready to learn from what one writer might do to his draft. Primarily, I was interested in cutting, in clearing away the underbrush and revealing the good writing hidden by it.

I was aware of the danger that I would take over the piece and make his vision and his voice mine. I hoped he would be strong and wise enough to resist me.

In going over this for the book, I have not changed my editing, which was done quickly for purpose of instruction and in a spirit of play, "I wonder what would happen if I . . . . . . ?" I have also realized how limited my editing was and that is good. I did not want to overwhelm the writer. I would edit myself more severely, and if I were editing the piece for publication I also would have caught all sorts of things I let go.

Go back and edit Roger LePage's piece alone or with a partner. Try to edit so that you reveal *his* vision of the world and free *his* voice from the draft. Work slowly with pen in hand. When you are finished, compare your editing with mine.

THE DUMP

Roger LePage, Jr.

■════════■

(very somehow) makes it less

(great image: smoke)

(Too many flat statements—draw us in as you have below.)

~~Summer days in Rollinsford, New Hampshire were very boring until we discovered the dump. As ten-year-olds we didn't have summer jobs—except, of course, finding amusement and adventure. We didn't even do that well until the time one of us, maybe me, maybe one of the others, steered his bike off of Main Street and led the pack onto the dirt dump road.~~ The ~~intriguing~~ pillar of smoke that was perpetually rising above the trees from the Rollinsford, New Hampshire, dump may have drawn us in. Or it could have been the "NO TRESPASSING" sign and the locked gate—things like that always invited us into abandoned houses. ~~Whatever the reason,~~ we ~~discovered~~ the town dump ~~and were not bored for the rest of the summer.~~

we were 10-years-old, had a summer out of school, and where everything abandoned was, new, full ≡ of possibility.

That first day we didn't even make it into the actual dump. The dirt road leading in was about half a mile long, and about half way down that we discovered a slimy (wonderful word) little pond with ~~lots of~~ frogs and snakes around its fringe. We spent most of the day hunting, ~~them and~~ trying to catch them. ~~This didn't hold its appeal too long for me though.~~ But when ~~One of my friends,~~ Jason, ~~was very cruel and whenever he~~ caught a frog he would first put it on the road and torment it with a stick, ~~and~~ then he would pick it up and squeeze it to death in ~~his~~ one hand and throw the guts at whoever was within range. I was ~~made~~ the target twice and ~~decided it was time to go~~ took off for home.

another day

But ~~soon~~ we made it past ~~all the distractions along the road, which were many, including~~ the rusting corpses of several kitchen appliances and ~~one time there was~~ the ass end of a deer, ~~the stench~~ with a stench ~~of which was still strong~~ that followed us for a quarter mile ~~radius. But nothing could be compared to the adventures found in the dump itself. We began~~ But when we arrived at the dump we ~~in the~~ huge pile of discarded tires, immediately abandoning our bikes ~~right in front of it~~ and ~~trying~~ tried to race to the top. "King of Tire

Mountain" lasted for weeks, ~~producing many bruises and scuffles and many hurt feelings.~~ Then we built forts within the inner depths of Tire Mountain, ~~a much more peaceful game.~~ When it was time to try to furnish our forts we ~~began~~ explor~~ing~~ed the junk in the rest of the dump.

That was when we discovered ~~the greatest thing possible in a ten-year-old boy's mind:~~ dirty magazines. ~~I'm not sure who discovered them first, but there was soon a huge uproar and race to see who could collect and hoard the most.~~ We ~~would bring~~ lugged them back to our forts and ~~pile them up, going through each one with great enthusiasm. We would~~ pile them, organize them ~~according to their appeal to us:~~ the ones with girls showing everything spread out ~~and accessible were our favorites;~~ were first, the ones with boys and girls ~~were~~ second, ~~the~~ plain boob shots were last. We had another pile of 'girls with girls' which we didn't know how to take but spent the most time looking at. We ~~would~~ moved around ~~in funny ways, always moving around~~ to maximize that ~~great~~ new ache in our pants while looking at the magazines, and we all seemed to have the shakes.

(watch out for would)   * * *

I ~~would~~ also ~~go~~ went to the dump on Saturday mornings with my dad. ~~It was part of our Saturday morning errands:~~ the dump, the barbers, the grocery store, then church. Most of the time he didn't even have anything to throw away. It was a Saturday morning meeting place for the ~~true~~ Rollinsford men—~~those~~ who still called the town Salmon Falls, and ~~a group to which my dad belongs because he was~~ were born and raised there. It was also a place of refuge from their wives: Friday night was poker night (I could always hear the door slam from my bedroom when my dad stumbled in just before day-break), and on Saturday morning the men were red-eyed and 'in the dog house' as they put it.

~~But~~ my dad ~~would go~~ went to the dump mainly to pay his respects to the toothless old Greek~~, who I've only heard referred to as The Greek, and who was~~ a ~~great~~ drinking partner of my dad's father. The Greek ~~worked there. He~~ made sure you didn't put metals or combustibles in with the burnables and he took his job very seriously. He tended the fire and if you tried to interfere, his toothless smile would disappear and he would ~~straightway~~ banish you. Banishment from the dump is ~~one of~~ the ~~lowest~~ greatest dishonors in our town.

~~You wouldn't think anyone would be overly enthusiastic about being friends with the man who tends the dump, but in towns like mine, occupation is not nearly as socially relevant as age or drinking ability. There were other reasons why~~ [because/when] The Greek was respected; ~~he did once have an occupation, but~~ [(from what?)] [he didn't] he retired ~~and instead of sitting~~ around all day, or ~~bagging~~ groceries ~~in some~~ [at the] supermarket, ~~he opted~~ [but took] [control of] ~~to tend~~ the dump. ~~That choice was respectable and The Greek knew this and took his job very seriously. Also~~ [^] he was [a] great old drinker and fighter, actually (legendary.) His missing teeth were ~~the proof.~~ ~~Everybody knew that~~ "he got them" knocked out" in a fight years back. The important thing was that "he got them knocked out", not "somebody knocked his teeth out" because the latter would imply The Greek was not in control at the time. ~~And even though~~ it was rumored that his wife, also something of a drinker and fighter herself, has his tooth marks on her knuckles ~~The Greek was still respected;~~ ~~he~~ earned the right to be toothless. He brandished this honor too, by smiling all the time—but I never thought The Greek was ~~such~~ a happy man. [(Great – put all them in a deeper context – a stage of growing & growing away from a parent)]

After my friends and I started spending so much time in the dump I began to look at the Saturday morning visits with my dad in a new way. I fancied myself a spy, watching our forts as inconspicuously as possible and always listening for clues to find out if anyone was on to us. I thought that sometimes The Greek would catch me looking at the tire pile and possibly know our secret, but I wasn't sure. I felt that it was just strange that he looked at me at all, because before my friends and I started playing at the dump, The Greek never seemed to notice me. I thought then, that maybe I just never noticed him noticing me before because it wouldn't have mattered if he did. Whatever the case, I was wary.

* * *

~~Then~~ [when] one day we were playing at the dump, alternating between culling our dirty magazines and our next best discovery: combustibles. It was Jason, ~~the sadistic frog~~ [of course, who] ~~squisher who discovered this one, I'm sure; he was fond of destruction in any form. On one of his devilish whims, he~~ threw a paint can into the dump fire and ~~silently~~ waited for something to happen. The rest of us were busy going through junk and exploring and didn't even notice. The explosion sent us running for cover, yelling all of the profanities we knew, (most of which we got out of our new magazine collection) ~~and nearly pissing our pants.~~ When we realized what ~~had taken place~~, Jason had done, the excitement ~~of the whole thing~~ brought us together giggling, wrestling and rolling around in the dirt. ~~This was probably why~~ we didn't hear the pick-up coming down the dump road.

~~When we did finally notice it,~~ it ~~had~~ pulled up right beside us, and We ~~all~~ jumped to our feet, ~~and stood there,~~ not sure what to do. I recognized the green, rusty, beat up old thing (Plymoth, Ford, what?)—it was The Greek's truck. He got out and stared at us with ~~full blown adult~~ contempt, ~~"What are you kids doing?~~ "You the ones been down here vandalizing?"

Even at ten I wondered how the hell anybody could *be* vandalizing a dump, but ~~as I said,~~ The Greek took his job ~~very~~ seriously.

"Aren't you Roger's boy? I knew I'd catch you down here," he stared at me with his exaggerated toothless frown and shook his head. His nose looked like a spoiled piece of fruit. "There's nothing but piss and vinegar running through your veins. I know. Knew your grandfather, same thing. Your dad too," he was still staring and shaking his head, the loose flaps of his jaw skin shook around violently. "I knew I'd catch you down here. How would you like me to tell your old man what you've been doing?"

No ten-year-old wants his parents to know what he does on summer afternoons; ~~it's a child's first glimpse of independence and the privacy is treasured, locked away in the child's heart. When one comes along and tries to adulterate that treasure, as adults always do, the loss of the treasure is a far greater tragedy than any thought of impending punishment.~~ I asked him, ~~meekly,~~ "Please don't tell." The Greek looked at me and then at my companions. He turned and walked along the dirt towards Tire Mountain. ~~I thought I was going to be sick.~~ The Greek knew! He was on to us. First we would be banished from the dump, then from our homes and then we would be kicked out of the town altogether. Maybe even sent to prison or reform school.

"How would your mothers feel if they knew you were looking at those kinds of things?"

Oh God! I thought. ~~It's one thing to lose your cherished summer freedom, or even to be locked away, but to have your mother think you're a pervert,~~ to have ~~her~~ mother think you are ~~one of those weirdos~~ like fat, old Slow Jimmy who smiles at you funny and always asks you to go for a walk. To have your mother know you look at dirty magazines! ~~It was absolutely the worst thing imaginable.~~

The Greek paused and looked around. He looked at me hard. I'm sure he could see I was about to ~~commit the unspeakable in front of my friends—that I was about to~~ cry. ~~He had done his job, and done it well. The Greek took his job very seriously.~~ "Well, if I catch any of you kids here again, your parents will all get a call from me. And if I catch you here again LePage, I'll give you a beating myself."

~~So we pedaled the hell out of there, still shaken and guilty.~~ We didn't ~~resolved never to~~ return to the dump, ~~and we didn't . . . . at least~~ for ~~the rest of the~~ a week. ~~But after all, as kids we also took our jobs very seriously.~~

The Student's Reaction to Professional Editing

The good writing student will listen to what the readers of his drafts have to say—fellow writers, instructors, editors—but will resist when necessary. LePage reacted strongly to my line-by-line editing:

> *I'll start off with the problem I had with the revised story, just to get it out of the way. The first thing I did when I received your edited version of my draft was to rewrite the story using all of your suggestions. Then I read it aloud over and over and something just didn't sound right. I have no specific examples of the revisions which caused this, but some element seemed to be missing. I was down to the bare essentials of the story, the skeleton. The problem with this, as I see it, is all skeletons look alike. It's the flesh that gives the narration character.*
>
> *I'll use Raymond Carver as an example. He is sometimes referred to as a "minimalist," (as I'm sure you know,) because at first glance his stories appear to be whittled down to skeleton form. However, that is not completely the case. He adds the flesh to the narration where it is necessary and he does it in such a way that the narrator, whether first person or not,*

booms with personality. For example, in the excellent story, "Cathedral," Carver describes a meal in this way:

"We dug in. We ate everything there was to eat on the table. We ate like there was no tomorrow. We didn't talk. We ate. We scarfed. We grazed that table. We were into serious eating." I don't know any editors, but I would guess that most would pull their hair out over a passage like this. There is unnecessary repetition and clichés. But what's great about it is we are allowed to understand the narrator: he is a regular guy with a good sense of humor and a natural joy for life. We understand this only because Carver knows how and when to break the rules, and he knows how his narrator is supposed to sound.

So I had to make the story sound right to me. When I talk about sound, I think I'm talking about voice, but I'm not sure. My idea of sound is this: when you're sitting at your writing desk and reading your work aloud and the person in the next room doesn't ask what you're reading from, but who you're talking to, then it's good. I think this is the most important element in a story, that it sounds right. The second would be that you have a good story to tell.

What the revisions did was help to make my story move. I'm learning slowly, mostly as I gain confidence and experience, that I don't need to reinforce every point I have shown through anecdote or image by stating it explicitly. Phrases such as "a much more peaceful game," not only slow the story down but insult the reader. Other phrases like, "the greatest thing possible in a ten-year-old boy's mind," cause a similar problem. For one thing, I might be alienating some readers who may not have been, at ten years old, as enthusiastic to see dirty magazines as I was. Further, even if all ten-year-olds do share this interest, there is no need for me to state it. I'm getting between the reader and the story. I am inserting an annoying little voice that whispers over the reader's ear and tells him how to feel and what to think, just in case my narrator is not doing his job.

The way I originally opened the story was also a sort of distraction in that it bored the reader, rather than engaging his interest. Instead of flatly stating that we were bored and *the dump was exciting, it is much better to open with the same image that attracted us into the dump: the pillar of smoke. The reader will then pick up on our journey right at the exact point where we began and join us in discovering the source of the smoke. The reader needs to "see" what we saw, the actual scene, not some filtered, transmuted second-hand account of it. The idea of voice is still pertinent here, though it's much more subtle, almost subliminal. It's the idea that the reader becomes subconsciously aware of the person, the personality, telling the story and will pass judgment on that personality just as on any other stranger. Whether the reader likes or dislikes this personality is irrelevant; what matters is that the reader trusts him. The only way I know to make the reader trust my narration is to tell the truth, pimples and all. If a Freudian were to get a hold of this piece and make conclusions about me or my upbringing, so be it. I had to tell it like it is.*

I also had problems describing The Greek. Aside from the flat statement similar to those that opened the story, and aside from the annoying little voice that tried to dictate how the reader should feel about The Greek, I created a kind of refrain with, "The Greek took his job very seriously." The technique could work, I believe, but only if the story was centrally about this character. My story did not center around The Greek but around a group of ten-year-old boys and for this reason a refrain about The Greek is misleading.

Lastly, my concluding paragraph was originally too drawn out, and like the opening paragraph, too flat. The fact is, we were afraid, embarrassed and anxious to flee, but all that should be obvious. Again, I don't need to state it. The only problem with the revised ending was that it didn't sound right. It was too abrupt. I had to find a medium point that kept the rhythm of the piece, sounded right, yet didn't give the reader the urge to skip over it.

My worry now is that in correcting old mistakes I have created new ones. I don't know if I'll ever in my life be able to say, "Okay, this is done."

That is an ideal example of a student response. He pays attention and learns but makes up his mind. It is, after all, his essay.

The Student's Revision

THE DUMP

by Roger LePage, Jr.

■══════════■

The pillar of smoke that perpetually rose above the trees into a puffy gray cloud over the Rollinsford, New Hampshire dump may have drawn us in. Or it could have been the "NO TRESPASSING" sign and the locked gate—things like that always invited us into abandoned houses. As ten-year-olds out of school for the summer we only wanted to be where we shouldn't or where the possibility of adventure seemed fullest. The town dump was to become our spot.

The first day we didn't even make it into the actual dump. The dirt road leading in was about half a mile long, and about halfway down that we discovered a slimy little pond with lots of frogs and snakes around its fringe. We spent most of the day hunting, trying to catch them. But that eventually stopped being fun, at least for me, because when Jason caught a frog he would first put it on the road and torment it with a stick, then he would pick up the frog and squeeze it to death in one hand and throw the guts at whoever was in range. I was the target twice and took off for home.

But another day we made it past all the rusting corpses of stoves, refrigerators, and washer machines and even past the ass end of a deer with a stench that seemed to linger around us for at least a quarter mile. When we got into the dump, we first came upon the huge pile of discarded tires and immediately ditched our bikes right in front of it and raced to the top. "King of Tire Mountain" lasted for weeks, and each day when we became too bruised and tired for that game, we built forts within the inner depths of the mountain. When it was time to try to furnish our forts we headed for the junk in the rest of the dump.

That was when we discovered dirty magazines. There was a huge race to see who could collect and hoard the most. We hurried them back to our forts and then organized them: the ones with girls showing everything spread out was the first pile, the ones with boys and girls was second, plain boob shots were last. We had another pile of 'girls with girls' which we didn't know how to take but spent the most time looking at. We lay down on our stomachs to look at them, maximizing that new ache in our pants, and we all seemed to have the shakes.

* * * *

I also went to the dump on Saturday mornings with my dad: the dump, the barbers, the grocery store, then church. Most of the time he didn't even have anything to throw away. It was a Saturday morning meeting place for the Rollinsford men—those who still called the town Salmon Falls and were born and raised there. It was also a place of refuge from their wives: Friday night was poker night. (I could always hear the door slam from my bedroom when my dad stumbled in just before day break,) and on Saturday morning the men were red-eyed and 'in the dog house' as they put it.

My dad went to the dump mainly to pay his respects to the toothless old Greek, a drinking partner of my dad's father. The Greek made sure that you didn't put metals or combustibles in with the burnables and he took his job seriously. He tended the fire and if you tried to interfere, his toothless smile would disappear and he would banish you. Banishment from the dump is the greatest dishonor in our town.

The Greek was respected because when he retired from the railroad he didn't sit around all day or bag groceries at the supermarket. He took over the dump. Also he was a legendary old drinker and fighter. His missing teeth were "knocked out" in a fight years back. The important thing was that "he got them knocked out," not "somebody knocked his teeth out" because the latter would imply that The Greek was not in control at the time. It was rumored that his wife, also something of a drinker and fighter herself, had his tooth marks on her knuckles. But he earned the right to be toothless. And he brandished this honor too, by smiling all the time—though I never thought for a minute The Greek was a happy man.

After my friends and I started spending so much time in the dump I began to look at the Saturday morning visits with my dad in a new way. I was no longer a mere tag-along, I had a need to be there of my own—a mission. I was not a child by my father's side, but a grown man, a spy watching our forts as inconspicuously as possible and always listening for clues to find out if anyone was on to us. I thought that sometimes The Greek would catch me looking at the tire pile and possibly know our secret, but I wasn't sure. I felt that it was just strange that he looked at me at all, because before my friends and I started playing at the dump The Greek never seemed to notice me. This, I decided, was because of my new identity: a man, (even if he is a spy,) automatically earns another's notice. But just in case, I remained wary.

* * *

One day when we were playing at the dump, alternating between culling our dirty magazines and exploring the junk piles, we came upon our next best discovery: combustibles. It was Jason, of course, who threw a paint can into the dump fire and waited for something to happen. The rest of us were too engrossed in the junk to notice. The explosion sent us running for cover, yelling all of the profanities we knew—most of which we got out of our new magazines collection. When we realized what Jason had done, the excitement heaved us together giggling, wrestling and rolling around in the dirt. We didn't hear the pick-up coming down the dump road.

It pulled up right beside me and we jumped to our feet, not sure what to do. I recognized the green, rusty, beat up old "Ford"—it was The Greek's truck. He got out and stared at us with contempt. "You the ones been down here vandalizing?"

Even at ten I wondered how the hell anybody could vandalize a dump. The Greek did take his job seriously.

"Aren't you Roger's boy? I knew I'd catch you down here," he stared at me with his exaggerated toothless frown and shook his head. His nose looked like a spoiled piece of fruit. "There's nothing but piss and vinegar running through your veins. I know. Knew your grandfather, same thing. Your dad too." He was still staring and shaking his head, the loose flaps of his jaw skin shook around violently. "I knew I'd catch you down here. How would you like me to tell your old man what you've been doing?"

No ten-year-old wants his parents to know what he does on summer afternoons. I asked him, "Please don't tell."

The Greek looked at me and my companions. He turned and walked along the dirt towards Tire Mountain. The Greek knew! He was on to us. First we would be banished from the dump, then from our homes and then we would be kicked out of the town altogether. Maybe even sent to prison or reform school.

"How would your mothers' feel if they knew you were looking at those things?"

Oh God! I thought. To have your mother think you're like fat, old Slow Jimmy who smiles at you funny and always asks you to go for a walk. To have her know you look at dirty magazines!

The Greek paused and looked around. He looked at me hard. I'm sure he could see I was about to cry. "If I catch any of you kids here again, your parents will all get a call from me. And if I catch you here again LePage, I'll give you a beating myself."

We scrambled for our bikes never to return again, and we didn't—for the rest of the week anyway.

▪ DOES REVISION EVER END? ▪

No. Revision is a game the writer plays for life. Roger LePage writes, *"I don't know if I'll ever in my life be able to say 'Okay, this is done.'"* And years earlier the French poet Paul Valery said, "A poem is never finished, only abandoned."

I heard despair in those quotations when I was as young as Roger LePage, but I now feel excitement and opportunity. This edition of *The Craft of Revision* is greatly different from the last one. I don't know if it is better—I hope so—but it is different. After more than fifty years of re-writing, I have learned from this draft. It is finished—abandoned—only because of the deadline. Writing inspires writing and I treasure the fact that each morning as I come to my writing desk, mystery and surprise await. My bones may creak, but as a writer I am forever young, forever learning as I write and re-write.

AFTERWORD

▪ RE-WRITE YOURSELF ▪

We become what we write. As we draft and revise, we become exposed to ourselves, and when our writing is read by others we become exposed to them.

> *I think after a while a writer can begin to know himself through his language. He sees someone or something reflected back at him from these constructions. Over the years it's possible for a writer to shape himself as a human being through the language he uses. I think written language, fiction, goes that deep. He not only sees himself but begins to make himself or remake himself.* DON DELILLO

> *Writers are not just people who sit down and write. Everytime you compose a book your composition of yourself is at risk. You put yourself further away from whatever is comfortable to you or you feel at home with. Writing is a lifetime act of self-displacement.* E. L. DOCTOROW

> *I never know what my stories are about until they are finished, until they choose to reveal themselves. I merely feel their power, how they breathe on me. I try not to write them. I prefer the rush of having them write me.* KATE BRAVERMAN

At first we have an idea, perhaps only the hint of an idea, a hunch, a passing thought, an unexpected feeling, a fleeting memory, an assignment that is unclear, a fragment of language, a problem without an apparent solution, an image, as yet unconnected facts and statistics, something overheard, something read and puzzled over.

As we re-write, we focus on what started us writing, collect information, shape and order it, find an appropriate voice and edit what we have written. The result reveals what we think and feel about the subject.

The stories we tell about our childhood, schools, marriages, wars, beliefs become who we are. On the basis of our narratives, told and re-told, we take our place in the family and the office, the community and the world.

Writing is the most effective way to use language to define our lives and to come to terms with the problems we confront.

Bharati Mukherjee recently said, "When my writing is going well, I know that I'm writing out of my personal obsessions." We all are. Our obsession may be the effective management of a super-market, profitable investment in real estate, careful placement of children up for adoption, accurate reporting of an election, making a police report that will stand up in court, writing a eulogy to be read at a friend's funeral, creating a sales letter, writing a screenplay or a computer software manual that is user effective. All the forms of writing tell how we think and how we feel.

Do not fear this. Writing and re-writing is therapeutic as the novelist Graham Greene points out:

> Writing is a form of therapy; sometimes I wonder how all those who do not write, compose or paint can manage to escape the madness, the melancholia, the panic fear which is inherent in the human situation.

Writing and re-writing gives us an opportunity to confront the major issues in our academic, professional and private lives. It allows us to revise our lives by understanding the world in which we live and our role in it. We can re-write ourselves.

Index